# Lateral Strain: Key to Detonation Control

Parkar

# Table of Contents

# Chapter 1

# Introduction

## 1.1 Motivation

Detonation waves are self-sustained supersonic combustion waves [1]. These waves are led by a shock, which compresses the fresh reactive media to a much higher temperature and pressure for rapid reaction [1]. The tremendous reaction heat release occurring behind the shock in return energizes the propagation process. As such, this closely coupled shock-reaction complex self-sustains. Detonation waves can be sustained in a variety of energetic media including reactive gases. The large overpressures generated behind gaseous detonations make them attractive and useful for developing propulsion systems [2], such as rotating detonation engines (RDEs) [3, 4] and pulse detonation engines (PDEs) [5, 6]. These applications require reliable control of the accurate ignition and stable propagation of a detonation wave. Likewise, for safety applications [7, 8], it is also desirable to have the predictability for the eventual initiation of a detonation wave and for its propagation limits when different mitigation strategies are used [9]. Therefore, realizing all these practical purposes requires predictive capability of detonation behavior.

Detonations in gases usually propagate with lateral strain. For example, in confined geometries of small size, such as narrow channels or tubes, detonations are subject to significant losses induced by boundary layers, which act as a mass sink and result in flow divergence in reaction zones, thereby giving rise to lateral strain impacting the detonation propagation [10]; while in geometries of varying cross-section areas or curved channels, as typically seen in PDE pre-detonator tubes and RDE combustors, detonations are curved with the flow also diverging after passing the leading front [11–13]. These lateral strain rates are generally known to decrease the detonation speed and its propagation limit [10, 11, 13–17]. Thus, in order to achieve the practical purposes of either utilizing or avoiding detonations, the effect of such lateral strain rates on detonation dynamics, e.g., detonation propagation speed and its failure limit, needs to be quantified for revealing their influencing mechanism. This, however, is still poorly understood and not well treated, since the highly unsteady nature of the well-known multi-dimensional structures of detonations greatly complicate the efforts.

In the present study, the objective is to contribute to the understanding of the effect of lateral strain rates on real gaseous detonation dynamics from both experiments and numerical modelling, and in particular, to carefully address the question in a well-posed manner whether the macro-scale dynamics of the complicated transient multi-dimensional detonations in real gases can be predicted by simple first principles models.

As the inherent unsteadiness and instability of gaseous detonations pose a considerable difficulty in determining the relationships between lateral strain rates and the detonation dynamics [18, 19], it necessitates a well-posed approach in tackling this challenge. The very recent work of Radulescu and Borzou [17] proposed a novel experimental technique for addressing these issues. It involves an exponentially diverging channel, which enables detonations to propagate with a constant mean lateral strain rate in quasi-steady conditions at the macro scale [17]. Its advantage, in precisely quantifying the effect of lateral strain rates on detonation speeds and propagation limits, permits meaningful comparisons between the analytical modelling results and experimental observations to be made in an unambiguous way. The present work thus employs this exponential horn technique for performing a systematic study of the effect of lateral strain rates on dynamics of detonations in a wide range of mixtures.

On the other hand, it is well known that detonations in narrow channels and tubes experience significant lateral strain due to boundary layers, which can considerably decrease the propagation speed as well as the propagation limit [10, 15, 16, 20, 21]. Although great efforts have been made on modelling these effects in one-dimensional (1D) detonations [10, 11, 21–23], their influence on the multi-dimensional structure remains unknown, which typically requires direct Navier-Stokes calculations resolving the boundary layers. Such a direct attempt, however, is computationally prohibitive. In addressing this problem, the present study formulates an unsteady quasi-2D approach for evaluating the effect of the boundary-layer-induced lateral strain rate (from the third dimension) on the detonation structure. To the author's knowledge, this is the first attempt to model the realistic boundary layer effects in 2D detonations.

## 1.2  Background State-of-knowledge

### 1.2.1  Detonation structures

The first theory that quantifies the detonation phenomenon was formulated by Chapman and Jouguet [1]. They proposed the calculations of the detonation parameters, i.e., speed, pressure, temperature, and density, through a control-volume analysis of the detonation wave by applying the conservation laws to determine the outflow conditions based on the known initial state. Its propagation speed, relying exclusively on the net energy release of reactants via reactions, is the minimum steady speed of a detonation wave. Such theoretical speed is known as the Chapman-Jouguet (CJ) velocity [1], at which a detonation travelling is considered ideal. The CJ theory, however, provides no insight into the structure of detonations. The internal structure and propagation mechanism of an ideal detonation

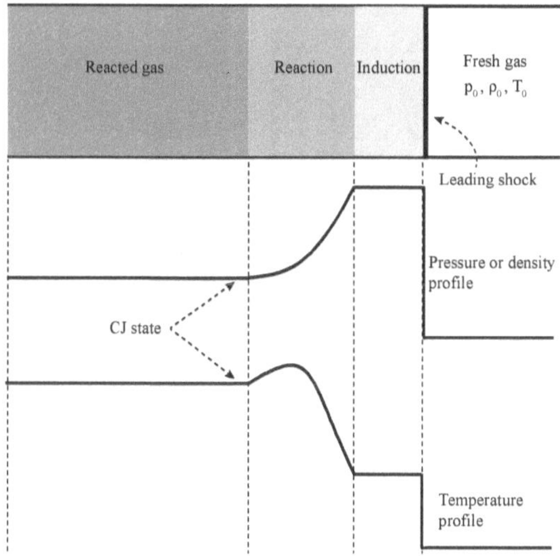



Figure 1.1: Illustration of a steady 1D ZND detonation wave, which travels from left to right, with pressure, density, and temperature profiles shown below.

wave were first described by the classical Zel'dovich-von Neumann-Döring (ZND) model [1], which treats a detonation as a steady inviscid 1D combustion wave. Figure 1.1 is a schematic diagram illustrating the structure of a ZND detonation wave. It comprises a supersonic leading shock (propagating at CJ speed) followed by a thermally neutral induction zone with the trailing reaction zone. The fresh reactive gases are compressed by the leading shock and then begin to react in the reaction zone after an induction delay. The expansion of the burned products in turn drives the shock front, thus forming a self-sustained coupled shock-reaction complex.

Despite these pioneering models of ideal steady detonations, real gaseous detonations have been well known to take on a transient multi-dimensional cellular structure, which consists of an intricate ensemble of interacting triple points, shear layers, and transverse waves [24–30]. For example, Fig. 1.2 shows the experimentally visualized cellular structures as well as the explanatory sketches of detonations in two typical mixtures, i.e., $2H_2/O_2/7Ar$ and $CH_4/2O_2$ [31]. The key feature of a detonation is the triple point, which intersects the incident shock (IS), Mach stem (MS), and the downstream transverse shock (TS) [1]. Emanating from the triple point is also the shear layer, which separates the reacted gases processed by the stronger Mach stem from the unburned ones by the weaker incident shock. The roles of incident shocks and Mach stems alternate along the leading front with the collision of triple points (including their collision with the wall). Depending on the strength, the transverse shock can be either reactive in burning the gases closely behind it or non-reactive [32]. When tracing the track of triple points along the leading shock, one can obtain the fish-scale-like cells [33]. The maximum spacing between the paired triple

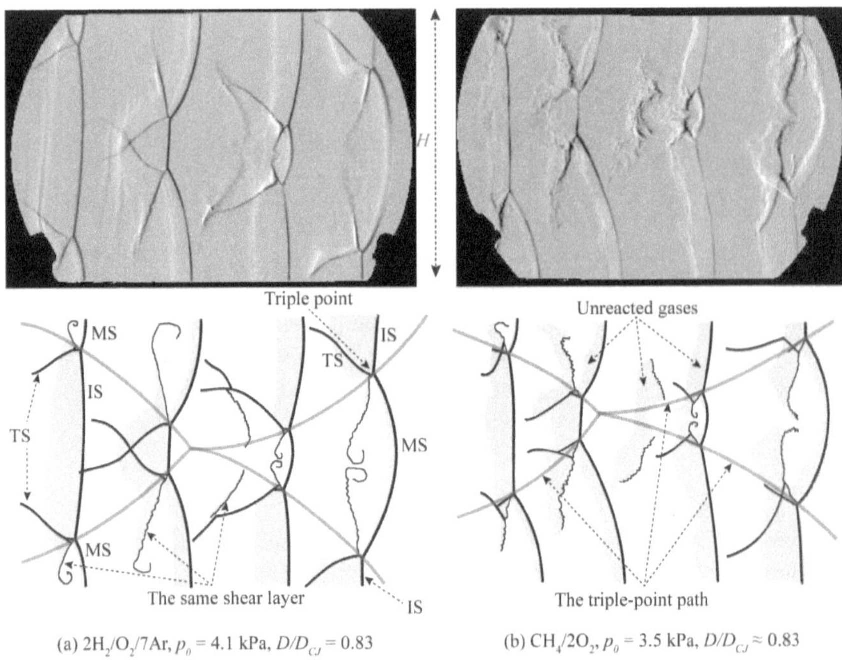



(a) $2H_2/O_2/7Ar$, $p_0 = 4.1$ kPa, $D/D_{CJ} = 0.83$    (b) $CH_4/2O_2$, $p_0 = 3.5$ kPa, $D/D_{CJ} \approx 0.83$

Figure 1.2: Evolution of experimentally visualized detonation cellular structures in different mixtures in a 2D channel. Detonations propagate from left to right. The top portions show the superimposed schlieren photographs while the bottom portions are the corresponding explanatory sketches. The schlieren photos of $CH_4/2O_2$ detonations are adapted from Bhattacharjee's work [31]. IS: Incident shock, MS: Mach Stem, TS: Transverse shock. The shock tube channel height $H$ is 203 mm and the channel width is 19 mm.

points is the characteristic detonation cell width $\lambda$, an important length scale characterizing the detonability of a mixture [1].

Depending on the regularity of cell patterns, detonations are generally classified as having either a *regular* cellular structure or an *irregular* structure [28, 34]. Figure 1.3 shows, for example, both the regular and irregular detonation cells obtained experimentally by Austin [28] using the soot-foil technique [34]. Clearly, the irregular detonations exhibit much more stochastic-looking and variable cellular patterns with sub-cell structures, whose cell size is difficult to define and measure. Such cellular patterns are also an indication of the detonation front instability [28], which has been found to correlate well with the effective activation energy ($E_a/RT_s$, $R$ the universal gas constant, $T_s$ the post-shock temperature) multiplied by the ratio of ignition time ($t_{ig}$) to reaction time scale ($t_{re}$). This was proposed by Radulescu [35] for quantifying the instability through the

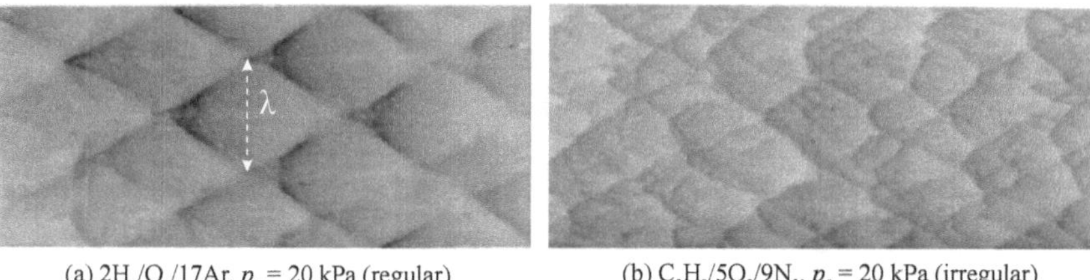



(a) $2H_2/O_2/17Ar$, $p_0$ = 20 kPa (regular)  (b) $C_3H_8/5O_2/9N_2$, $p_0$ = 20 kPa (irregular)

Figure 1.3: Cells of (a) regular detonations and (b) irregular detonations at the initial pressure of 20 kPa [28]. Note that the channel height of the shock tube, used by Austin [28], was about 152 mm.

definition of $\chi = (E_a/RT_s) \times (t_{ig}/t_{re})$. Regular detonations with a much smaller $\chi$ value (empirically less than 10 [35, 36]) appear to be much less unstable than the irregular ones in their structures. Typical regular mixtures are the ones highly diluted with a monotonic gas such as argon. They are characteristic of low activation energies, which result in very limited variations of ignition delays for the shocked reactive mixtures behind the wave front structure. The principle ignition mechanism for regular detonations is believed to be via adiabatic shock compression [1, 37, 38]. Moreover, the reaction zone structures of regular detonations tend to be piecewise-laminar, as can be seen from Fig. 1.2a with little signature of turbulence. However, for irregular detonations in hydrocarbon-oxygen mixtures as shown in Fig. 1.2b, the reaction zone structures are highly turbulent [37, 39, 40]. Compared to regular detonations, the shocked gases behind irregular detonations have a much larger distribution of ignition delays, due to higher activation energies of the irregular mixtures. Another distinctive feature of irregular detonations is the formation of significant unreacted gas pockets or islands, whose consumption is mainly through diffusive flames facilitated by turbulent mixing [37, 39, 40] instead of being done by shock compression alone. These differences presumably give rise to the distinct macroscopic behaviors in the propagation and failure mechanism of regular and irregular detonations [1, 17, 35].

## 1.2.2 Detonations in small tubes and narrow channels

Gaseous detonations in small tubes and narrow channels have been experimentally measured to propagate at a smaller mean speed than the ideal CJ detonation speed, with large velocity deficits due to wall losses to the side walls [10, 11, 21–23, 41, 42]. For example, in the narrow channel experiments of Fig. 1.2, the average detonation velocity deficit can reach approximately 20% of the CJ speed. In addition to the velocity deficit, the critical pressures $p_c$ [15, 16, 20, 21, 34, 35, 43, 44], below which detonations fail to propagate, and the characteristic cell sizes [34, 45, 46] are also increased for detonations in these small-size confinements. Thus, the first interesting question is whether these macro-scale detonation dynamics can be predicted by simple first principles models, in spite of the time-varying multi-dimensional cellular structures of detonations.

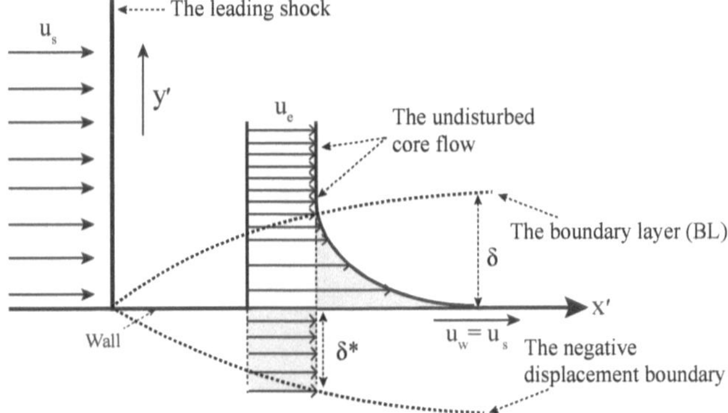



Figure 1.4: Sketch of the boundary layer (BL) velocity profile behind the leading shock in the shock-attached frame of reference. $u_s$ is the leading shock speed in the laboratory frame, while $u_e$ is the post-shock flow velocity outside the boundary layer in the shock frame. $u_w$ is the wall velocity in the shock frame. $\delta$ and $\delta^*$ represent the boundary layer thickness and displacement thickness, respectively.

In 1950, Zel'dovich [22] was the first to model the wall losses on impacting detonation propagation velocity and detonability limits, by incorporating the heat and friction losses into the conservation laws for 1D steady detonations. This is the first extension of the classical ZND model. This model was later reviewed in detail and developed by Gelfand et al. [11], Agafonov and Frolov [43], and Kitano et al. [17]. Zel'dovich-type generalized ZND model assumed the heat and friction losses to be uniformly distributed throughout the whole cross section. It was later pointed out that such a physical picture was questionable since the effects of heat transfer and momentum losses due to the non-slip wall were only confined to the boundary layer [10]. Their influence was communicated to the core flow only through pressure gradients in the subsonic reaction zone [10]. Fay [10] then proposed a more realistic mechanism of boundary layer effects. Figure 1.4 illustrates the schematic diagram of the boundary layer velocity profile behind the leading shock, in the shock-attached frame of reference. Since the flow in the boundary layer has a larger velocity than the undisturbed main stream, it appears that some fluid is "leaking" through the boundary layer from the core flow. Thus, the boundary layer (in the shock-attached frame of reference) acts as a sink by removing the mass from the main stream flow [10]. Equivalent to the core flow, the tube seems to have an enlarged cross section, whose area increase can be determined by the boundary layer displacement thickness $\delta^*$ that allows to accommodate the excess of mass flow "leaked" through the boundary layer [10]. As a result, a flow divergence in the reaction zone behind the detonation front is induced with a negative boundary layer displacement, thereby reducing the propagation velocity.

In the work [10], Fay analytically derived a reduced formula for evaluating the velocity deficit. Gooderum's turbulent boundary layer thickness relation [48] was adopted as the displacement thickness $\delta^*$ for evaluating the boundary-layer-induced flow divergence or equivalently the lateral strain rate. A reasonable agreement was reported between experiments and the Fay model predictions. One should note that Fay's conclusion of turbulent boundary layer behind detonations is questionable, since Liu et al. [49], Damazo et al. [50], Chinnayya et al. [51], and Sow et al. [52] have demonstrated that the boundary layer behind detonations is most likely laminar. Fay's velocity deficit relation was further extended by Murray [53] and subsequently applied in recent works on detonations in small tubes and narrow channels [46, 54–56]. The researchers have found that, the experimentally obtained detonation velocity deficits and propagation limits are in generally good agreement with the model predictions for weakly unstable detonations with regular cellular structures; while for the highly unstable detonations with irregular cells, the agreement is poor. Nevertheless, these Fay-type models require a number of simplifying assumptions and matching constants [17]. For example, a specifically defined value of pressure ratio $\epsilon$ and a particular length scale for evaluating the flow divergence are essential for these models. The impact of these assumptions on the model predictions has not been evaluated, which may diminish the value of the comparisons.

The other more refined extension of Fay's model was carried out by Dove et al. [21]. Fay's turbulent boundary layer displacement thickness relation was still employed for evaluating the flow divergence involved in the governing equations for the steady detonation model. This generalized ZND model with lateral strain rates was characterized by a set of ordinary differential equations (ODEs) and can be solved numerically. A detailed reaction mechanism was used for describing the chemical reaction kinetics. This extended ZND model was then popularly applied in predicting dynamics of detonations in small tubes and narrow channels, e.g., see the experimental works of Chao et al. [20] and Camargo et al. [44]. The concept of lateral flow divergence was also applied to model the dynamics of detonations propagating in porous tubes [16, 35, 57]. In their works, a constant flow divergence was assumed, which was estimated from the permeability of the porous wall by assuming a choked flow [35]. Again, qualitatively good agreement was obtained for regular detonations while large deviations for irregular detonations were observed. In all these works, however, limitations exist in the unrealistic assumption of a uniform or constant flow divergence in porous and narrow tubes, as demonstrated by Chinnayya et al. [51] and Mazaheri et al. [57]. This has been clearly argued at some length in the recent work of Radulescu and Borzou [17]. All these factors thus potentially impact the conclusions made in these works.

It is thus clear that, although great efforts have been made in modelling the dynamics of real gaseous detonations in small-sized confined geometries, the question regarding the predictability of detonation dynamics by a priori models is still not well addressed. The ambiguity of their conclusions results from some controversial factors, as briefly discussed above. Therefore, it requires another more well-posed approach in answering this question.

### 1.2.3 Curved detonations

Detonations are usually curved adapting to changes in confinement geometry. Even in the periodic cellular structure, the Mach stem also takes on the form of a curved detonation, as

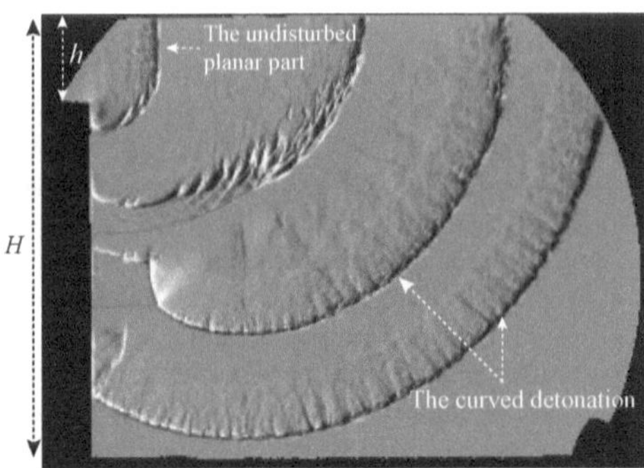



Figure 1.5: Evolution of curved detonations subject to abrupt area changes, in the mixture of $2H_2/O_2/2Ar$ at the initial pressure of 23.5 kPa. The entrance channel height $h$ is 38 mm, while the whole cross-section area height $H$ is 203 mm.

can be easily seen from Fig. 1.2. As a result of the curvature, flow across the leading shock is diverged in the trailing reaction zone, thus giving rise to flow divergence or equivalently the lateral strain inside the detonation. Moreover, the curvature of a detonation front can be quantitatively correlated with its induced lateral strain rate in a simple mathematical expression [12, 13]. Therefore, the effect of curvature is essentially identical to that of lateral strain on detonations.

The theoretical studies [13, 58, 59] have demonstrated that, for a specific front curvature $\kappa$, or equivalently the same lateral strain rate, there will be a unique detonation speed $D$ that can permit its post-shock flow to accelerate to the sonic condition [13]. Given a $D(\kappa)$ relation for a detonable mixture, the dynamics of detonations in complex geometries can be predicted by relevant models, for example in the framework of Detonation Shock Dynamics (DSD) [60].

While the $D(\kappa)$ curve can be obtained from the real chemistry calculations using the quasi-steady 1D detonation model [13], it can also be extracted from well-posed experiments. The question is *whether these theoretically predicted $D(\kappa)$ curves agree with those obtained from experiments*, which characterize the dynamics of real multi-dimensional cellular detonations. Compared to its popularity in condensed-phase explosives [61, 62], the $D(\kappa)$ theory is much less common in study of gaseous detonations, especially from experiments. Such scarcity results from the complication due to the notorious unsteadiness and cellular instability inherent in gaseous detonations. As shown recently by Jackson et al. [18, 19], local cellular pulsations can introduce considerably large local and temporal variations in normal detonation velocity $D$ and its curvature $\kappa$. These large spans can swamp any steady curvature and thus complicate determination of the $D(\kappa)$ relationship from the front shape measurement in regular experiments [18].

Curved detonations can be observed in the popular scenario of detonation diffraction [63–66]. When a planar detonation diffracts into a much larger cross-section area, as shown in Fig. 1.5 for the diffraction of $2H_2/O_2/2Ar$ detonations, the expansion waves emanating at

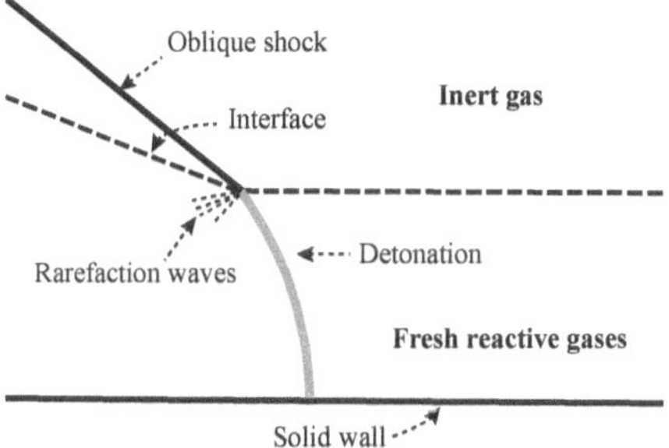


Figure 1.6: Sketch of a weakly confined detonation, which propagates from left to right.

the corner penetrate into the reaction zone and gradually disturb all the detonation front. As a result, curvature is introduced along the detonation front, with the flow diverging in the trailing reaction zone. Significant unsteadiness is also induced [64], which complicates the quantification of the relationship characterizing the detonation dynamics with lateral strain rates.

Another scenario of curved detonations happens in a reactive charge bounded by a yielding confinement, such as an inert gas, as shown in Fig. 1.6. This curved detonation phenomenon commonly occurs in RDEs [67], where the reactants are weakly confined by the burned products. Works on this topic mainly start from multi-dimensional simulations, for example, see Refs. [68–70]. In these works, they evaluated the lateral strain rate due to the inert confinement by adopting the methodology proposed by Wood and Kirkwood [11], and comparisons were made between the simulation results and predictions from the steady quasi-1D model of Wood and Kirkwood, in terms of velocity deficits with respect to lateral strain rate. They found that this model can well predict the multi-dimensional simulation results in mixtures of low activation energy; while for mixtures of higher activation energy, the model underpredicts the velocity deficits from simulations. This trend differs from the experimental observations, where the opposite trend was found [17,71]. Such a paradox has been clarified recently by Radulescu [71] as a result of diffusive effects, which were absent in these inviscid simulations. Despite these simulation works, corresponding well-posed experiments are required for more convincing insights.

Recently, Kudo et al. [72] and Nakayama et al. [73,74] adopted arc channels for experimentally investigating the dynamics of curved detonations. Figure 1.7 shows, for example, the sketch illustrating the curved detonation evolution process in an arc channel, corresponding to the experiments of Nakayama et al. [73,74]. After a relaxation stage from the initial planar one, the curved detonation achieves a quasi-steady state. Nakayama et al. [73,74] then spatially evaluated the local curvature of these quasi-steady curved detonation fronts as well as their local speeds, and finally established the $D(\kappa)$ relationships. They were the first to apply the $D(\kappa)$ concept to characterize the dynamics of gaseous detonations from the experimental perspective. Moreover, they found a universal dimensionless $D(\kappa)$ relation for the three tested mixtures.

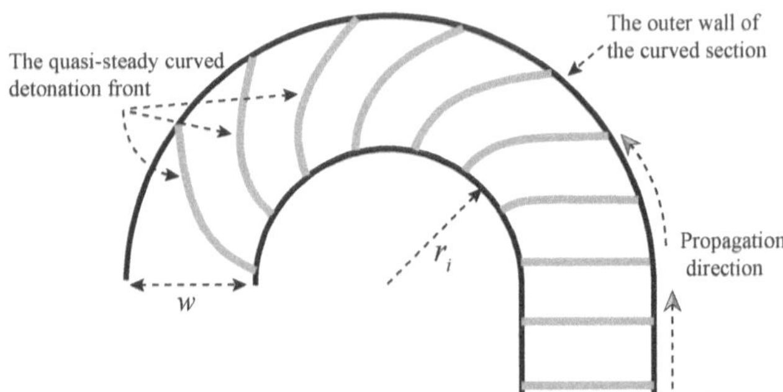



Figure 1.7: Sketch of the curved detonation evolution in an arc channel. In experiments of Kudo et al. [72] and Nakayama et al. [73,74], the fixed channel width $w$ is 20 mm, while the inner radius $r_i$ can be varied from 5 mm to 60 mm. Note that their channel depth was either 1 mm [73] or 16 mm [72,74].

More recently, Radulescu and Borzou [17] designed a pair of exponentially diverging ramps (exponential horns) for inducing specifically curved detonations, as shown in Fig. 1.8. The key feature of this novel formulation is to keep the derivative of the logarithmic cross-section area with respect to distance as a constant. Such property of maintaining a constant logarithmic area divergence rate enables detonations inside the channel to propagate in quasi-steady conditions with a constant mean lateral strain rate, i.e., an equivalently constant mean front curvature [17]. As such, a mean $D(\kappa)$ relation can be inferred from the experiments for characterizing the detonation dynamics with the lateral strain rate, allowing for making meaningful and unambiguous comparisons between experiments and relevant theories. In their work [17], two mixtures of different regularity were tested, i.e., the highly unstable one of $C_3H_8/5O_2$ and weakly unstable one of $2C_2H_2/5O_2/21Ar$. Firstly, they demonstrated that detonations in the well-posed exponential horns propagated at a constant average speed, which was controlled by the magnitude of the lateral strain. Beyond a critical value of lateral strain rate, detonations were not possible. Moreover, they compared the experimentally obtained $D(\kappa)$ curves with those predicted by the generalized ZND model in the presence of lateral strain rate. They found that the predictions made with the extended steady ZND model disagreed with the experiments. The less unstable $2C_2H_2/5O_2/21Ar$ detonations showed better agreement between experiments and predictions than the more unstable detonations in $C_3H_8/5O_2$. The question then arises that, can the extended ZND model better predict detonation dynamics of much less unstable mixtures, in spite of the cellular structures? Also, is there a link of departure between experiments and the ZND model predictions to detonation instability? Answering these questions thus becomes the objective of the present study.

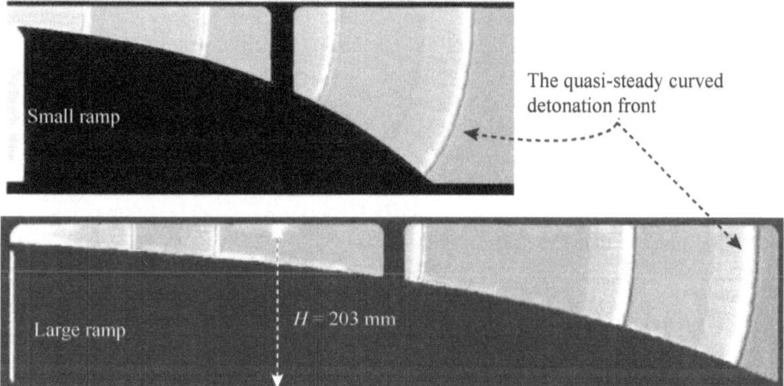



Figure 1.8: Evolution of $C_3H_8/5O_2$ detonations in exponentially diverging channels (exponential horns), adapted from the work of Radulescu and Borzou [17]. The initial pressure of these experiments was 12.1 kPa [17]. The channel width or depth $w$ was 19 mm. Detonations propagate from left to right.

### 1.2.4 Summary

In light of the review provided in this chapter, the outstanding scientific questions regarding detonations with lateral strain rates can be summarized as:

- Can the extended ZND model with lateral strain rates better predict detonation dynamics of much less unstable mixtures such as the $H_2/O_2/Ar$ reactive system, as compared to the ones studied by Radulescu and Borzou [17]? If so, then why?

- What is the effect of lateral strain rate on unstable detonations in other hydrocarbon-oxygen mixtures, such as methane-oxygen ($CH_4/2O_2$), ethylene-oxygen ($C_2H_4/3O_2$), and ethane-oxygen ($C_2H_6/3.5O_2$)? Is there a link of departure between experiments and the theoretical predictions to detonation instability? How can we improve and make corrections to the ZND model predictions of the unstable hydrocarbon-oxygen detonation dynamics?

- What is the effect of lateral strain rate on the unsteady 2D detonation cellular structures in narrow channels? The fact, that the answer to this question still remains unknown, is because: (a) from the experimental perspective, it is very hard to isolate the lateral strain rate for evaluating its effect on the unsteady 2D detonation cellular dynamics; (b) computationally, the already developed models are mainly on 1D detonations, and no efforts have been made in modelling realistically the effects of the lateral strain rate on unsteady 2D cellular detonations in narrow channels.

Therefore, the present work aims to address the above three questions. For the first one and second one, this **book** will extend the exponential horn technique of Radulescu and

Borzou [17] to the very regular $H_2/O_2/Ar$ reactive system, and also other hydrocarbon-oxygen mixtures with varied levels of detonation instability. For the third question, since directly resolving the boundary layer in 3D simulations by solving the NS equations is computationally prohibitive, this work will for the first time propose a novel quasi-2D formulation of the reactive Euler equations, accounting for the boundary-layer-induced loss from the third dimension in narrow channels. Complementary experiments in a corresponding narrow channel will be performed for validating the model. With this quasi-2D approach, the effect of the boundary-layer-induced lateral strain rate on the unsteady 2D cellular detonations can thus be computed.

## 1.3  Outline of This Study

The present **book** reports the work based on three articles. The body of this work comprises two parts.

The first part applies the well-posed exponential horn technique to a range of mixtures, from the very regular one such as $2H_2/O_2/7Ar$ to the highly irregular one of $CH_4/2O_2$, for experimentally investigating the detailed dynamics of detonations with a constant mean lateral strain rate. In this part, two papers are included.

The second part is a formulation of a numerical model to evaluate the effect of boundary layer losses on the cellular structures of 2D detonations. Complementary experiments in a narrow channel are reported for validating the model. This part includes one article.

# Chapter 2

# Experimental Investigation on Dynamics of Detonations with a Constant Mean Lateral Strain Rate

## 2.1   Overview

This chapter aims to quantify the effect of lateral strain rate on detonation dynamics, and moreover, to explore the predictability of these macro-scale detonation dynamics by the generalized ZND model. The well-posed exponential horn technique of Radulescu and Borzou [17] will be extended to a range of characteristic mixtures with varied levels of detonation instability, for experimentally investigating the detailed dynamics of detonations with a constant mean lateral strain rate. Two articles are included, as shown respectively in sections 2.2 and 2.3.

The first paper, in section 2.2, focuses on very regular detonations in the reactive $H_2/O_2/Ar$ system with varied argon dilutions, i.e., $2H_2/O_2/2Ar$, $3Ar$, $4.5Ar$ and $7Ar$. Experiments are performed in a wide range of initial pressures. Propagation characteristics of these detonations inside the exponential geometry are demonstrated in detail. Furthermore, comprehensive comparisons between experiments and the generalized ZND model predictions of quasi-1D detonation dynamics with lateral strain rates are made. While for the second paper in section 2.3, it is motivated by experimentally obtaining the characteristic $D(\kappa)$ relationships for detonations in various hydrocarbon-oxygen mixtures with varied cellular instability, including stoichiometric methane-oxygen ($CH_4/2O_2$), ethylene-oxygen ($C_2H_4/3O_2$), and ethane-oxygen ($C_2H_6/3.5O_2$). Based on such a wide range of mixtures, whether there is a link of departure between the experimentally obtained detonation dynamics and the generalized ZND model predictions to detonation instability is established.

## 2.2 Dynamics of Hydrogen-Oxygen-Argon Cellular Detonations with a Constant Mean Lateral Strain Rate

**Status:** This paper has been published in *Combustion and Flame*. The preliminary work was first presented at the 69th Annual Meeting of the APS Division of Fluid Dynamics in Portland, USA in 2016, and the subsequent progress was presented at the 26th International Colloquium on the Dynamics of Explosions and Reactive Systems (ICDERS) in Boston, USA in 2017 [75].

This study revisits the problem of modelling the real gaseous detonation dynamics at the macro scale by simple steady 1D models. Experiments of very regular cellular detonations were performed inside the exponentially diverging channels (exponential horns), in the reactive $H_2/O_2/Ar$ system with varied argon dilutions (ranging from 40% to 70% in volume fraction). Characteristic reaction zone structures of relatively low initial pressure detonations were reported in detail. The paper showed that quasi-steady detonations can be obtained at the macro scale, with a constant mean lateral strain rate. As a result, meaningful relations between the mean propagation speeds and lateral strain rates were established. In an unambiguous manner, excellent agreement was found between experiments and the generalized ZND model predictions of quasi-1D detonation dynamics with lateral strain rates, using the detailed chemical kinetics. A novel clarification for this excellent agreement was proposed. Moreover, the Mirels' laminar boundary layer theory was, for the first time, applied to evaluate the boundary-layer-induced lateral strain rate. Good agreement between experiments and the predictions of Mirels' model further adds support to the hypothesis of laminar boundary layer behind detonations.

# Dynamics of Hydrogen-Oxygen-Argon Cellular Detonations with a Constant Mean Lateral Strain Rate

## 1. Introduction

Real detonations in gases have been experimentally observed to travel at a speed smaller than the ideal Chapman-Jouguet (CJ) detonation speed by a velocity deficit, due to the presence of non-ideal effects [1, 2]. These non-ideal factors include lateral flow divergence, unsteadiness, and momentum and heat losses [1]. Extensive efforts have been made to quantitatively compare the experimentally measured velocity deficits and propagation limits with the theoretical predictions made by relatively simple models, which build up on the classical one-dimensional (1D) Zeldovich-von Neumann-Doering (ZND) model [3–16]. The multi-dimensional transient cellular structures, consisting of an intricate ensemble of interacting triple points, shear layers, and transverse waves, of the real gaseous detonations, however, greatly complicate these attempts.

They give rise to substantial deviations from the classical 1D ZND detonation structures [2, 17]. A significant question hence comes up that, can the extended ZND model, which neglects the time varying cellular structures, be able to model the real detonation dynamics in the presence of losses at the macro-scale? The present paper addresses this question.

This question has also been attempted in the past by investigating detonation propagation in narrow tubes [8, 9, 11–15] and tubes with porous walls [18, 19]. The mean propagation velocities of detonations under varied initial pressures as well as the propagation limits were experimentally determined for various mixtures. Due to the lateral flow divergence from the growth of the viscous boundary layer on tube walls acting as a mass sink [4] or from the porosity of the walls, streamlines in the steady reaction zone are diverged resulting in a globally curved detonation front experiencing a velocity deficit [20]. In narrow tubes, the boundary layer theory of Fay [4] was adopted for evaluating the global lateral strain rate; while in porous tubes, where constant flow divergence was assumed, it was estimated from the permeability of the porous wall by assuming a choked flow [19]. The generalized ZND model with lateral strain was then applied to model the detonation dynamics. The authors have found that, the experimentally obtained detonation velocity deficits and propagation limits are in generally good agreement with the theoretical predictions, made with the steady ZND model for weakly unstable detonations, which are characterized by regular cellular structures; while for the unstable detonations with irregular cells, the agreement is poor. Nevertheless, a number of simplifying assumptions and matching constants were made in these works for the predictions. Firstly, there exist limitations in the unrealistic assumption of uniform flow divergence for detonations in narrow and porous tubes, as Chinnayya et al. [20] and Mazaheri et al. [21] have numerically demonstrated that a curved detonation front with flow divergence due to wall boundary layers or permeability is not expected to have a unique curvature. Moreover, Fay-type models [8, 12–15] require the empirical inputs of a specifically defined value of pressure ratio $\epsilon$, and a particular length scale for modelling the flow divergence rate, whose impact on the predictions has not been evaluated. All these factors thus potentially diminish the values of the comparisons in relevant works.

Very recently, Radulescu and Borzou [22] experimentally formulated a novel solution allowing for making a meaningful comparison of the experimental results with theoretical models. Their experimental technique involved two exponentially shaped channels. The constant logarithmic derivative of the cross-sectional area enabled detonations to propagate with a constant mean front curvature in quasi-steady state at the macro scale. Two mixtures of different regularity were tested, i.e., the highly unstable one of $C_3H_8/5O_2$ and weakly unstable one of $2C_2H_2/5O_2/21Ar$. Firstly, they showed that detonations in the exponential channels propagated at a constant average speed, which was controlled by the magnitude of the lateral strain. Beyond a critical value of lateral strain, detonations were not possible. Moreover, they compared the experimentally obtained relationship, between the detonation velocity deficit and its front's global curvature, with the generalized ZND model in the presence of lateral strain rate. The predictions made with the steady ZND model for the velocity deficit disagreed with the experiments. The less unstable $2C_2H_2/5O_2/21Ar$ detonations showed better agreement between experiments and the theoretical predictions than the more unstable detonations in $C_3H_8/5O_2$. An interesting question then arises that, can the extended ZND model better predict detonation dynamics of much less unstable mixtures, in spite of the cellular structures? This becomes the objective of the present work.

It is well known that argon-diluted $2H_2/O_2$ detonations have the weakest one dimensional (1D) instability among those typically investigated experimentally [17, 19] and have a very regular cellular structure. The chemical kinetics of $H_2$ decomposition is also better known than for hydrocarbons. Therefore, the present study aims to extend the above well-posed technique to more stable mixtures of $2H_2/O_2/2.0Ar$, 3.0Ar, 4.5Ar, and 7.0Ar, for the purpose of investigating in detail the dynamics of very regular cellular detonations with a constant mean lateral strain rate in exponentially diverging channels.

## 2. Experimental Details

The experiments were conducted in a 3.4-m-long aluminium rectangular channel with an internal height and width of 203 mm and 19 mm, respectively. A sketch of the experimental set-up is shown in Fig. 1, which is the same as that adopted by Radulescu and Borzou [22]. The shock tube comprises three sections, a detonation initiation section, a propagation section, and a test section. The mixture was ignited in the first section by a high voltage igniter (HVI), which could store up to 1000 J with the deposition time of 2 $\mu$s. Mesh wires were inserted in the initiation section for promoting the formation of detonations. Eight high frequency piezoelectric PCB pressure sensors (p1-p8) were mounted flush on the top wall of the shock tube to record pressure signals and then obtain the propagation speeds by using the time-of-arrival method. The test section was equipped with two glass panels in order to visualize the detonation evolution process. For the safety purpose of performing experiments at high initial pressures, the visualization glass panels were alternatively replaced by aluminum ones.

Two different polyvinyl-chloride (PVC) ramps, which enabled the cross-sectional area $A(x)$ of the channel to diverge exponentially with a constant logarithmic area divergence rate ($K = \frac{d(lnA(x))}{dx}$), were adopted in the test section. Dimensions of the ramps are shown in Fig. 1b. The large ramp had the logarithmic area divergence rate of 2.17 $m^{-1}$, while for the small one such rate was 4.34 $m^{-1}$. At the entrance, a protruded rounded tip was kept for minimizing the effects of shock reflection on the detonation front. The initial gap between the ramp tip and the top wall of the channel is 23 mm in height. The height between the exponentially curved wall and the top wall is given by $y_{wall} = y_0 e^{Kx}$ for $x > 0$ and by $y_{wall} = y_0$ otherwise.

The mixtures presently studied were stoichiometric hydrogen/oxygen with different argon dilution, i.e., $2H_2/O_2/2.0Ar$, 3.0Ar, 4.5Ar, and 7.0Ar. Each of them was prepared in a separate mixing tank by the method of partial pressures and was then left to mix for more than 24 hours. The mixture was introduced into the shock tube through both ends of the tube at the desired initial pressure with an accuracy of 70 Pa. Before filling with the test mixture in every single experiment, the shock tube was evacuated below the absolute pressure of 90 Pa. A driver gas of $C_2H_4/3O_2$ separated by a diaphragm, as shown in Fig. 1a, was used in the initiation section for low initial pressures, under which detonations of the test gas cannot be initiated successfully before entering the test section. For visualizing the detonation evolution process along the exponential ramp, a large-scale shadowgraph system was adopted by using a 2m×2m retro-reflective screen with an incandescent filament light source of 1600 W Xenon arc lamp from Newport. The resolution of the high-speed camera was 1152×256 $px^2$ with the frame rate of 42049 fps or 42496 fps. The exposure time was set to 0.81 $\mu$s. Alternatively, a Z-type schlieren setup [23] with a vertical knife edge was also utilized with a light source of 360 W. The resolution of the high-speed camera was 384×288 $px^2$ with the framing rate of 77481 fps and the exposure time of 0.44 $\mu$s. Note that the background of each shadowgraph and schlieren photograph in the present study was appropriately removed and the images were post-processed.

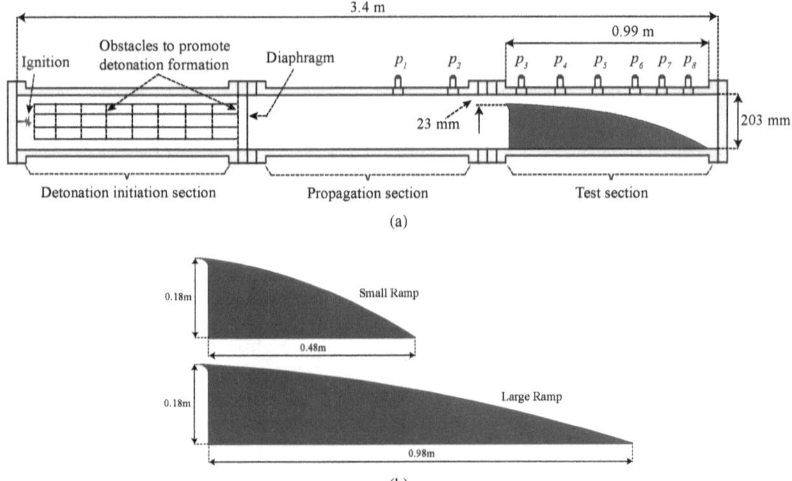

Figure 1: Experimental set-up for diverging detonation experiments: (a) the shock tube with the large ramp inserted in the test section and (b) the exponentially diverging ramps.

## 3. Experimental Results

### 3.1. Propagation of $2H_2/O_2/2Ar$ detonations along the large ramp

The superimposed shadowgraph photos illustrating the evolution of diverging detonation fronts along the large ramp for the mixture of $2H_2/O_2/2Ar$, at an initial pressure of 14.8 kPa and 12.4 kPa, respectively, are shown in Fig. 2. The detonation propagated from left towards right. The detonation front acquired a large number of small-sized cellular structures, and was notice-ably curved with a characteristic curvature due to the cross-sectional area divergence. Within the limited resolution of the photographs, transverse waves can be recognized starting from triple points and extending backward downstream. One can also note that, as the cross section area of the channel increases, new transverse waves were continuously generated and the average transverse wave spacing appeared to remain constant. This suggests that the cell size remains constant in the self-sustained propagation of diverging detonations along the ramp. The average detonation cell size measured in Fig. 2a and b is approximately 17.0 mm and 21.0 mm, respec-tively, which are found to be larger than the values of 10.5 mm and 13.5 mm obtained from the Detonation Database [24] for the same initial pressures.

On the other hand, the theoretically expected arcs of curvature from the quasi-1D approxi-mation, whose radius equals the reciprocal of the logarithmic area divergence rate $K = 2.17\text{m}^{-1}$, were obtained and compared with the real detonation fronts in experiments. These arcs of circles with the radius of $1/K = 0.46\text{m}$ are denoted by the dashed red lines in Fig. 2. The comparison shows that the detonation front's global curvature is in very good agreement with that expected by the quasi-1D approximation for designing the exponential geometry, despite some minor de-viations near the end of the ramp. These small deviations, as a result of the error in the quasi-1D

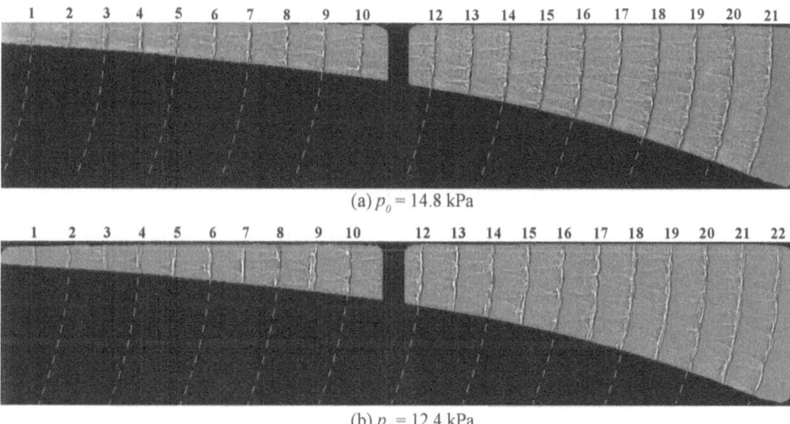

(a) $p_0 = 14.8$ kPa

(b) $p_0 = 12.4$ kPa

Figure 2: The superposition of detonation fronts along the large ramp at different instants for the mixture of $2H_2/O_2/2Ar$; red lines denote arcs of circles with the expected curvature from the quasi-1D approximation.

assumption in designing the exponential geometry, can be evaluated by [22]

$$\frac{K_{2D}}{K} = \left[1 + (Ky_{\text{wall}})^2\right]^{-1/2} \tag{1}$$

where the effective curvature $K_{2D}$ represents the curvature of the arc of a circle perpendicularly intersecting both the exponential wall at the point $(x, y_{\text{wall}})$ and the top wall as well.

Figure 3 shows the evolution of the mean curvature of detonation fronts shown in Fig. 2 as well as for detonations in other mixtures from Borzou's experiments [25]. The front curvature was fitted by using the least-squares method provided by SciPy. Results in Fig. 3 clearly show that it takes a relaxation length scale for the initially planar detonation front (before entering the ramp) evolving into the curved one due to the exponential geometry. For the large ramp, the relaxation length is approximately 0.4 m, while approximately 0.2 m for the small one. After the relaxation stage, detonations follow the evolution of the effective curvature $K_{2D}$ and can be reasonably assumed to obey a constant mean curvature in their propagation, in spite of the slightly decreasing curvature recognized near the end of the ramps. It thus suggests the appropriateness of the macro-scale quasi-1D approximation for detonations propagating inside these exponential channels.

The detonation propagation process for other lower initial pressures, near the limit, is illustrated in the superimposed shadowgraph photos of Fig. 4. With the decrease of the initial pressure for reducing the kinetic sensitivity of mixtures, detonation cells are considerably enlarged. The triple-shock structure, comprising a Mach stem, an incident shock, and a transverse wave that extends behind the detonation front, can now be clearly observed. One can also observe the consumption of the unreacted induction zones behind the incident shock by the passage of transverse waves, e.g., see Frame 15 through Frame 18 in Fig. 4b, implying the reactivity of these waves. This type of reactive transverse wave has been investigated first in detail by Subbotin [26] for the

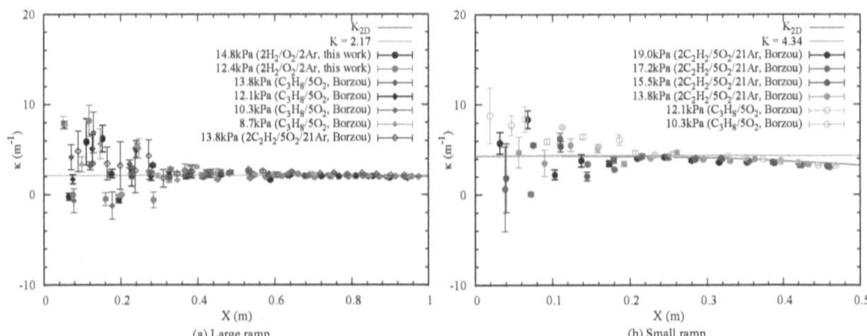

Figure 3: The mean curvature of the experimentally obtained curved detonation fronts at different instants. The green line is the constant curvature expected from the quasi-1D approximation while the red line denotes the effective curvature $K_{2D}$ of the geometry as shown in Eq. (1).

marginal, weakly unstable detonations and subsequently reported in a series of numerical and experimental works, e.g., Sharpe [27], Pintgen et al. [28], and Austin [29]. When the initial pressure was further decreased to approach the failure conditions, below which detonations are not possible, the reaction zone structure becomes very clear (Fig. 4c-e). These near-limit detonations were able to travel successfully with only one reactive transverse wave, which is interpreted as a transverse detonation [30], as can be easily seen from Fig. 4c to Fig. 4e. The main features of the transverse detonation, as shown clearly in Fig. 5, are qualitatively similar to that documented numerically by Gamezo et al. [30]. The transverse detonation, propagating transversely along the large induction zone, is strong enough to burn almost all the material except for a thin tail in the vicinity of the triple point [30]. The generation mechanism of this thin non-reactive tail in the gap between the leading shock and the transverse detonation front has been clarified as a result of the low temperature of the mixture in an embedded double Mach reflection [30]. When delineating the track of the triple point, one can readily obtain the single-headed detonation cell, denoted by the red dashed line in Fig. 4c-e. Clearly, as the detonation propagated towards the end, the detonation cell size considerably increased. An alternative explanation of the enlarging cells is the stabilization mechanism proposed by Short et al. [31]. Clearly in these curved detonations, the growth of the detonation front's area may be higher than the intrinsic transverse instability growth rate, thus resulting in the single-headed detonation of continuously increasing cell size without birth of any new transverse waves. At these near limit conditions, significant departures between the real detonation fronts and the arcs of circles of expected curvature in Fig. 4a-c can also be observed.

The locally averaged speeds both along the top and bottom curved walls were calculated from the shadowgraph photos in Fig. 2 and Fig. 4, with the distance between every two neighboring frames divided by their time interval $\Delta t$, and their profiles are shown in Fig. 6. The speed was normalized by the ideal CJ detonation velocity calculated with the NASA chemical equilibrium code CEA [32]. For calculating the speed along the curved wall, the arc length $L_{12}$ of the wall

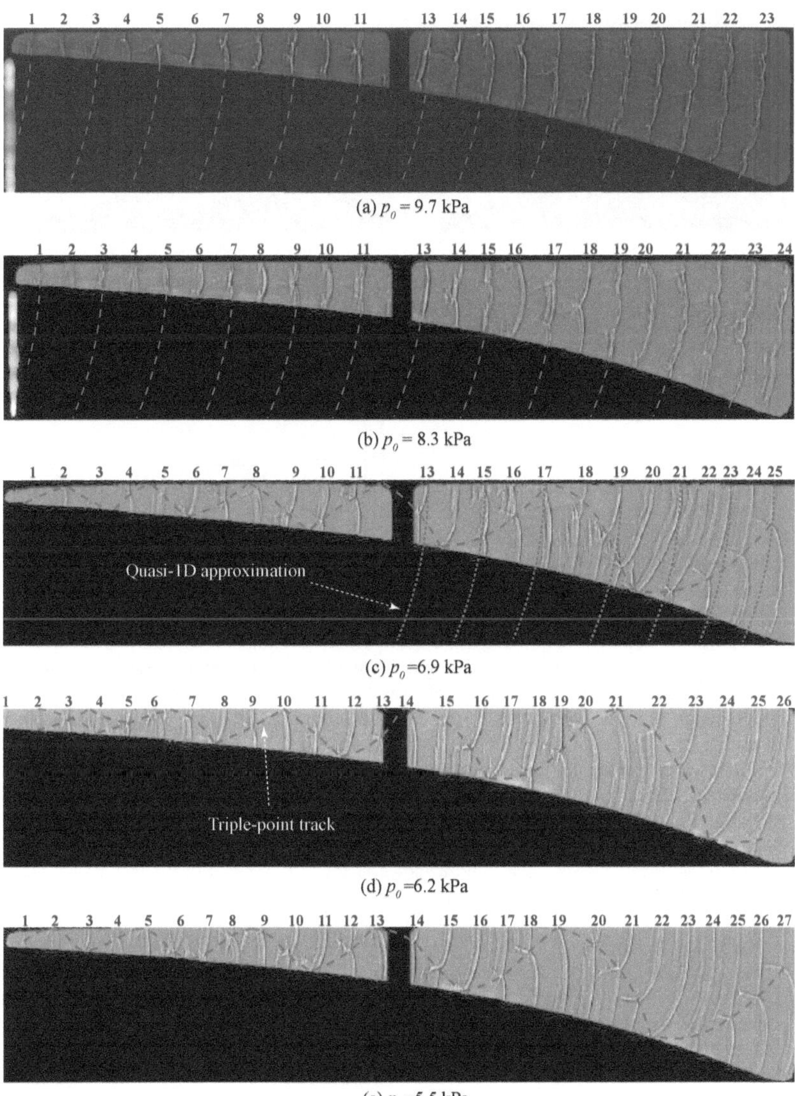



Figure 4: The evolution of $2H_2/O_2/2Ar$ detonation fronts along the large ramp at different initial pressures near the propagation limit.

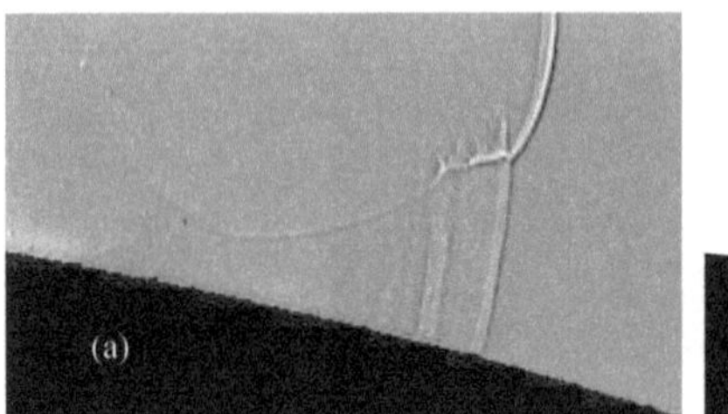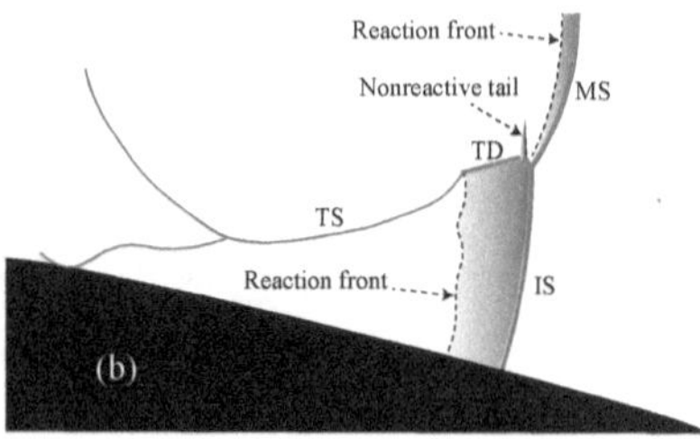


Figure 5: Details of the characteristic transverse detonation: (a) zoom-in of the cellular structure of Frame 21 in Fig. 4e, and (b) sketch of the main features. MS: Mach Stem, IS: Incident Shock, TD: Transverse Detonation, TS: Transverse Shock. Note that the shadow zone between the reaction front and the detonation front is the unburned induction zone.

segment between two points $(x_1, y_{\mathrm{wall}_1})$ and $(x_2, y_{\mathrm{wall}_2})$ can be evaluated by

$$L_{12} = \frac{1}{K}\left[\sqrt{1 + (Ky_{\mathrm{wall}})^2} + \ln\left(\frac{Ky_{\mathrm{wall}}}{1 + \sqrt{1 + (Ky_{\mathrm{wall}})^2}}\right)\right]_1^2 \tag{2}$$

The results in Fig. 6a-g demonstrate that within the current resolution, the local speed profiles show more significant variations for detonations at lower initial pressures. Especially for the critical detonations with one triple point, as shown in Fig. 6e-g, the local speed profiles exhibit periodic fluctuations, ranging from the maximum of 1.2 $D_{CJ}$ to the minimum of 0.5 $D_{CJ}$. These characteristic fluctuations very well illustrate the periodic evolution of the lead shock inside the cells. As the detonation propagated to the right end, the fluctuation period increased, implying an increasing length scale of the detonation cell. This is consistent with the finding from Fig. 4 that the detonation cell size increased as a result of the diverging area for the near-limit detonations. Noteworthy are the out-of-phase periodic velocity profiles measured along the top and bottom walls, shown in Fig. 6e-g. This can be interpreted as a result of the alternation between the strong Mach stem with relatively high speed and the weak incident shock of low velocity in the detonation front of a single triple point. The global mean propagation speeds along the top wall and bottom walls differed by at most 3% of $D_{CJ}$. Moreover, the running time average (of 5$\Delta t$) of the top-wall local speeds has also been shown in Fig. 6h. It again illustrates that for initial pressures well above the limit, detonations can be assumed to propagate in quasi-steady state at a macro scale much larger than the cellular structure with a constant speed (e.g., see the profiles of 14.8 kPa and 12.4 kPa).

For the critical pressure of 5.5 kPa, below which detonations were unable to propagate, six experiments in total were repeated. It was found that three of them successfully propagated as single-headed detonations while the rest three finally failed. This apparently stochastic behavior near the limits has also been observed in studies on detonation diffraction [33, 34]. Any slightly different perturbations during the whole evolution process can sensitively impact the result of Go or No-Go for detonation propagation in the critical pressure range. Figure 7 shows the failing process of the critical detonation and correspondingly its speed profiles. Before the failure, it was similar to that of the successful cases in the propagation mechanism of a single-headed detonation. As the detonation proceeded, the trailing reaction front was gradually detached from

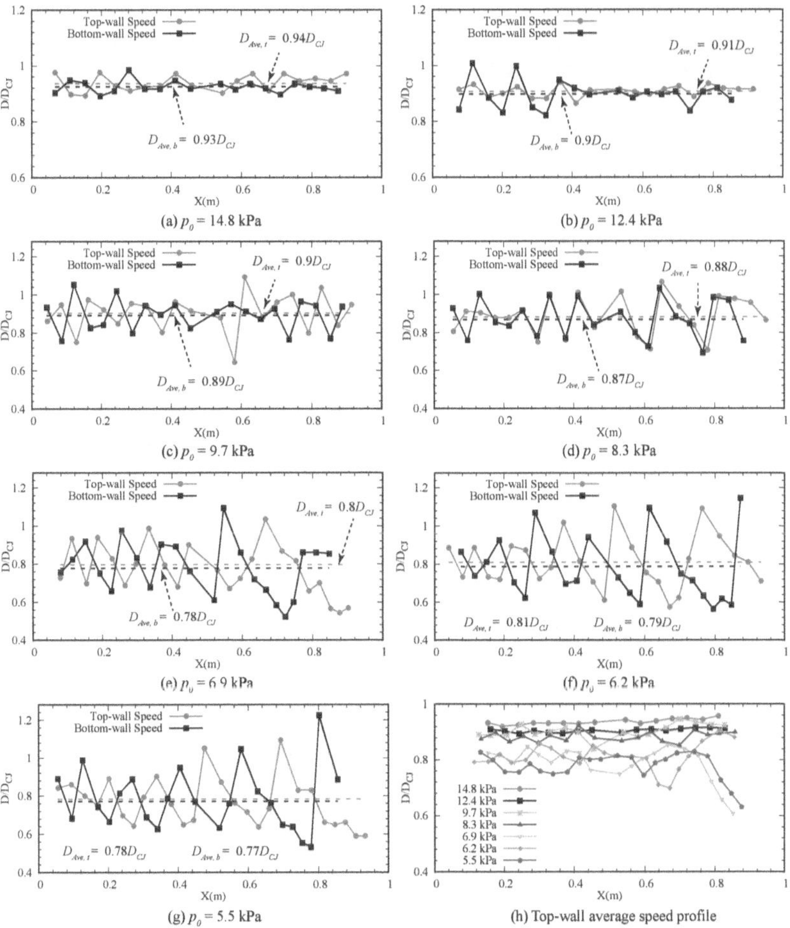

Figure 6: The local speed profiles (a)-(g) and the running time average ($5\Delta t$) of the top-wall local speeds (h), of detonations at different initial pressures in Fig. 2 and Fig. 4. The dashed lines represent the corresponding global mean propagation speed.

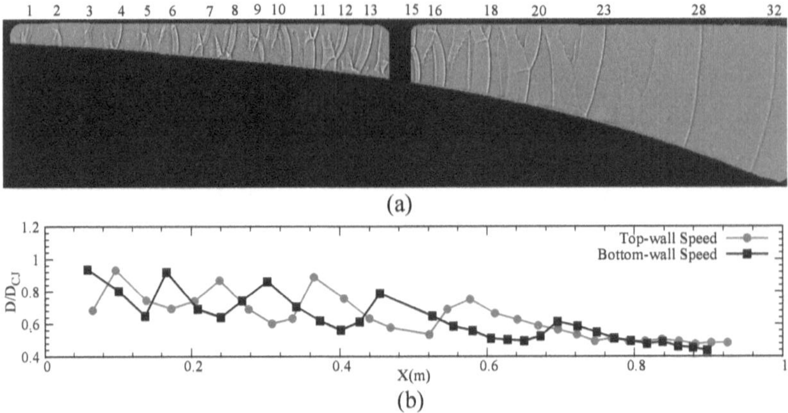

Figure 7: Shadowgraph photos (a) and local speed profiles (b) of a failed detonation at the initial pressure of 5.5 kPa for the mixture of $2H_2/O_2/2Ar$.

the leading shock and unreacted tongues were formed behind, with the transverse wave becoming inert. Finally, the detonation failed with the complete detachment of the reaction front from the leading shock (see Frame 28 in Fig. 7a), thereby resulting in the continuous decay of the shock, which can be seen from the continuously decreasing speed in Fig. 7b. Of noteworthy is that no distinctive transverse detonations were observed in the decoupled shock-flame complex for the failure case.

In addition, the average detonation cell size was also obtained from the shadowgraph photos of each experiment, as illustrated in Fig. 8. It characterizes the relationship between the ratio of the presently estimated cell size ($\lambda$) to previously reported cell size ($\lambda_0$) [35, 36], and the mean propagation speed, normalized by the ideal CJ value. Note that the relation of $\lambda_0$ was given by $\lambda_0(p) = 443.3p^{-1.39}$, which was fitted from the data in the Detonation Database [24]. Also, it is reasonable to assume these reported cell sizes for detonations with very limited or no velocity deficits, i.e., ideal CJ detonations. The results from Fig. 8 show that the normalized cell size ($\lambda/\lambda_0$) increases considerably when decreasing the mean propagation speed with respect to the ideal CJ speed ($D/D_{CJ}$), suggesting a strong dependence of the cell size on the detonation velocity deficit. This agrees with the finding of Ishii and Monwar [12] for detonations in narrow channels of varied sizes. When the mean propagation speed further decreased to $0.8D_{CJ}$, detonations started to be organized in a single-headed one, whose cell size was about 10~15 times larger than that of the ideal CJ detonation at the same initial pressure. Previous works proposed the onset of single-headed spinning detonations as a criterion for propagation limits of detonations in narrow tubes [37, 38]. Here, it further demonstrates the characterization of propagation limits inside the exponentially diverging channel by a single-headed detonation, organized with a distinctive transverse detonation.

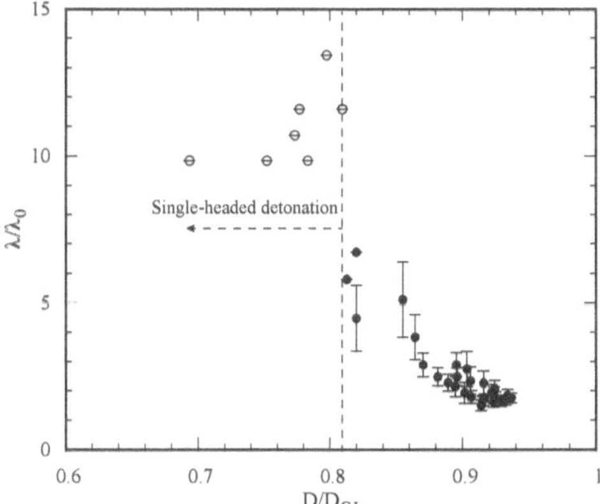


Figure 8: The ratio of estimated cell size ($\lambda$) in the present large ramp experiments to previously reported cell size ($\lambda_0$) as a function of the normalized mean propagation speed $D/D_{CJ}$ for detonations in the mixture of $2H_2/O_2/2Ar$.

### 3.2. Propagation of $2H_2/O_2/2Ar$ detonations along the small ramp

Experiments were also performed for detonations propagating along the small ramp of a higher area divergence rate of $K = 4.34m^{-1}$, i.e., double that of the large ramp. Visualization of the reaction zone structures of detonations at varied initial pressures near the limit, as shown in Fig. 9, enables us to make the following observations. At 10.3 kPa, detonations propagated with two triple points, i.e., one pair of transverse waves, along the detonation front (Fig. 9a). Evidently, these transverse shocks were reactive in burning the unburned gases behind the leading shock, preventing the generation of significant unreacted gas pockets [28, 29]. When the initial pressure was decreased to approach the critical one of 8.1 kPa, a single-headed detonation featuring a transverse detonation was formed, as is clearly illustrated in Fig. 9b and c. Below this critical pressure, the leading shock was found to continuously detach from the trailing reaction front with the transverse wave being inert, which can be seen from Fig. 9d. Since the average speeds along the walls were measured to be only 0.5~0.6 $D_{CJ}$, detonations can thus be interpreted as finally failed at the initial pressure of 7.9 kPa in Fig. 9d. It is not clear if such case along the small ramp with a much longer extended length in the stream-wise direction would have been possible.

The global mean propagation speeds of $2H_2/O_2/2Ar$ detonations, measured along the top wall in all the experiments of both ramps, are shown in Fig. 10 as a function of initial pressures. As a result of the flow divergence, experienced by detonations inside the cross-section area diverging channels, the mean propagation speed is smaller than its ideal CJ detonation speed by a velocity deficit. These velocity deficits increase with the rate of geometrical area divergence, as can be easily concluded in Fig. 10 from the higher propagation speeds measured in the large ramp experiments ($K = 2.17m^{-1}$) than that of the small ramp ($K = 4.34m^{-1}$) under the same initial pressures. As the initial pressure is reduced, the deviation of the mean propagation speed from the ideal CJ value becomes larger, implying a more significant role of area divergence in impacting detonations of lower initial pressures. Near the limit, such velocity deficits can reach 20% ~ 30%

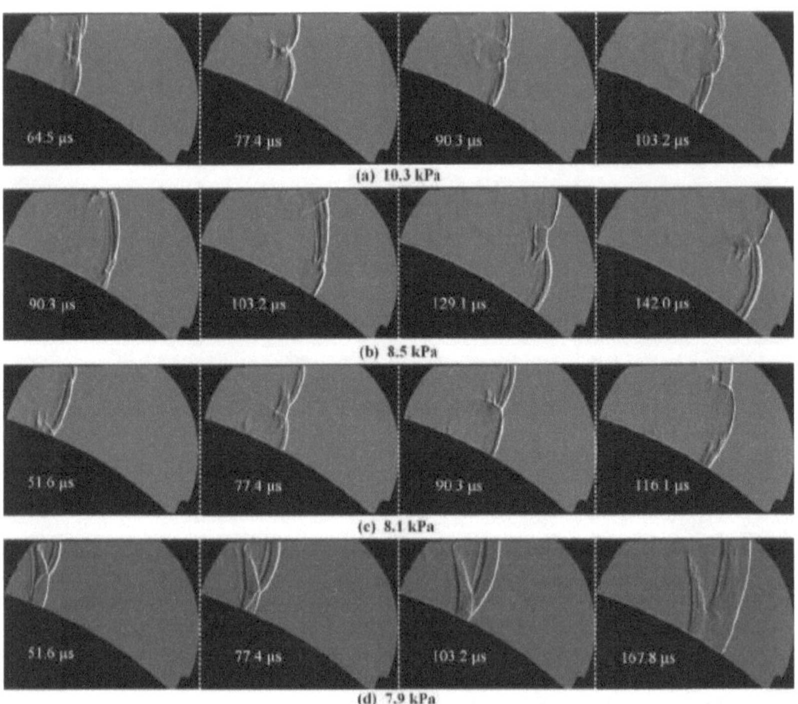



Figure 9: Schlieren photographs of near-limit detonations along the small ramp for the mixture of $2H_2/O_2/2Ar$.

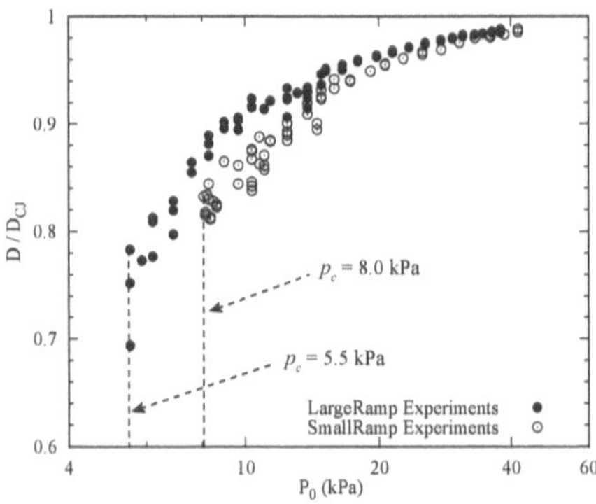



Figure 10: Average speeds normalized by the corresponding CJ speeds with respect to different initial pressures for the mixture of $2H_2/O_2/2.0Ar$.

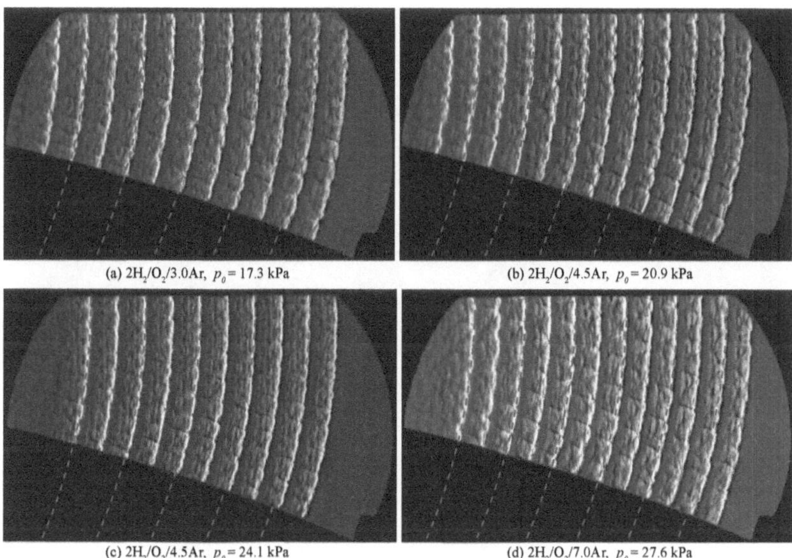

(a) $2H_2/O_2/3.0Ar$, $p_0 = 17.3$ kPa

(b) $2H_2/O_2/4.5Ar$, $p_0 = 20.9$ kPa

(c) $2H_2/O_2/4.5Ar$, $p_0 = 24.1$ kPa

(d) $2H_2/O_2/7.0Ar$, $p_0 = 27.6$ kPa

Figure 11: Comparison of the superimposed detonation fronts near the end of the large ramp with the arcs of curvature (denoted by red lines) expected from the quasi-1D approximation for different mixtures.

of the CJ value. One more observation is the higher propagation limit pressure $p_c$ of detonations along the small ramp. This can be interpreted as the consequence of more losses experienced by detonations along the small ramp due to its larger lateral strain rate, thus giving rise to a higher critical pressure.

### 3.3. Effect of argon dilution

In the present study, experiments of detonations in stoichiometric $H_2/O_2$ mixtures with other argon dilutions both along the large ramp and small ramp were also performed. It served the purpose of both qualitatively and quantitatively demonstrating the influence of argon dilution on detonation behaviors. The mixtures involved in this part are $2H_2/O_2/3.0Ar$, $4.5Ar$, and $7.0Ar$, corresponding to the argon dilution of 50%, 60%, and 70%, respectively. Figure 11 shows the superimposed schlieren photographs illustrating the evolution of detonations, well above the limit, along the large ramp for different mixtures. The curved detonation fronts were uniformly textured with triple points, with transverse waves extending downstream behind the leading shock. Again, cell sizes remain approximately constant in the exponentially diverging channel, under initial pressures far away from the limit. Moreover, comparisons of these experimentally obtained detonation fronts with arcs of the expected curvature from the quasi-1D approximation were also made. The very good agreement between the real curved detonation fronts and the theoretically expected arcs, denoted by the red dashed lines in Fig. 11, demonstrates the independence of the detonation front's global curvature on mixture compositions and initial pressures. It

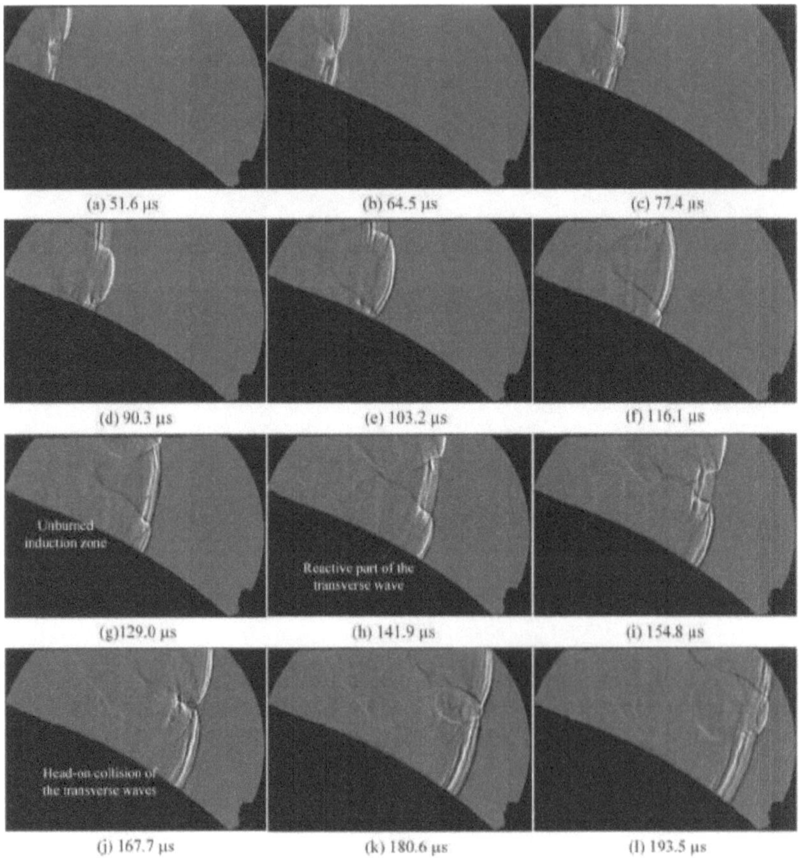



Figure 12: Schlieren photographs illustrating detonation evolution along the small ramp in the mixture of $2H_2/O_2/4.5Ar$ at the initial pressure of 11.2 kPa.

is thus indicative of the fact that argon diluted $H_2/O_2$ detonations, well above the limit, propagate in quasi-steady state at the macro-scale with a constant mean curvature.

The visualized cellular structures of near-limit detonations in mixtures of $2H_2/O_2/3.0Ar$, 4.5Ar, and 7.0Ar show qualitatively the same behaviors as that of $2H_2/O_2/2.0Ar$ detonations. For example, Fig. 12 gives the evolution process of detonations with one pair of triple-shock structures in the mixture of $2H_2/O_2/4.5Ar$ at the initial pressure of 11.2 kPa. It excellently demonstrates the interactions between the very regular cellular structures, including the collision of transverse waves and their reflections from the walls. Due to the reactive portion of the transverse shock, sweeping across the unburned induction zone behind the incident shock, no sig-

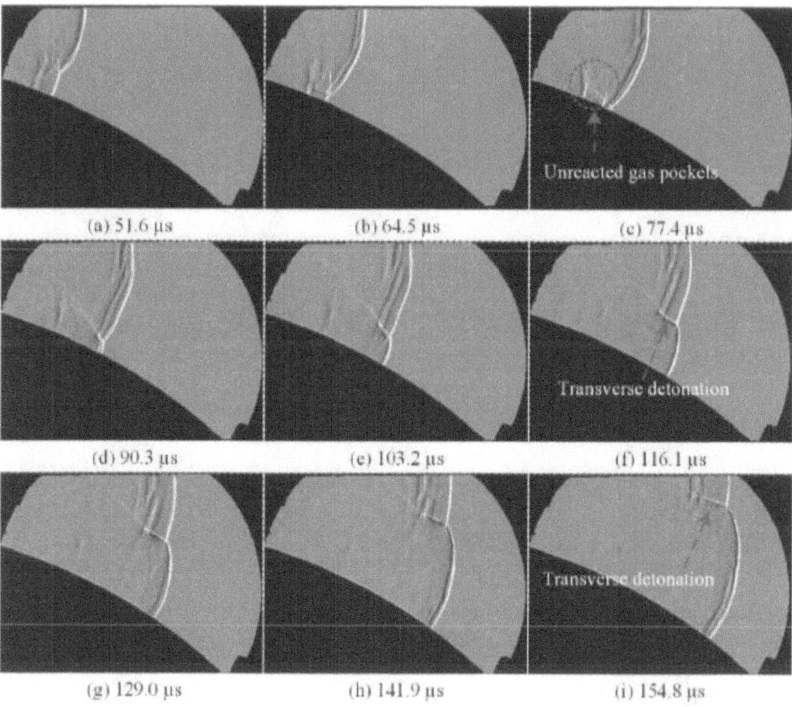


(a) 51.6 µs  (b) 64.5 µs  (c) 77.4 µs

(d) 90.3 µs  (e) 103.2 µs  (f) 116.1 µs

(g) 129.0 µs  (h) 141.9 µs  (i) 154.8 µs

Figure 13: Schlieren photographs illustrating detonation evolution along the small ramp in the mixture of $2H_2/O_2/3Ar$ at the initial pressure of 8.8 kPa.

nificant unreacted gas pockets were observed. The detailed propagation process of single-headed detonations is illustrated in Fig. 13 for the mixture of $2H_2/O_2/3.0Ar$. Prior to the formation of a transverse detonation, some unburned gases can be observed behind the transverse wave in the first three frames, implying that this transverse shock was non-reactive and of weak type. The transverse detonation occurred after the reflection of the inert transverse wave from the curved wall, which can be seen from the frames of 90.3 $\mu s$ through 116.1 $\mu s$ in Fig. 13. This behavior can be better observed in Fig. 14, showing the transition process between the non-reactive and reactive transverse waves in near-limit detonations in the mixture of $2H_2/O_2/7.0Ar$. The initially reactive transverse wave decayed gradually and transitioned to a non-reactive transverse wave, which could be seen from the frames of 64.5 $\mu s$ through 141.9 $\mu s$ (Fig. 14). After the reflection of the non-reactive transverse wave from the bottom curved wall, considerable unreacted gases were pinched off as pockets, e.g., see frames of 154.8 $\mu s$ and 167.7 $\mu s$. The reflected transverse wave again became reactive and rapidly transformed to a strong transverse detonation burning all of the unreacted gases behind the leading shock. After the generation of the transverse detonation, unreacted gas pockets were no longer produced. These observations confirm Subbotin's finding

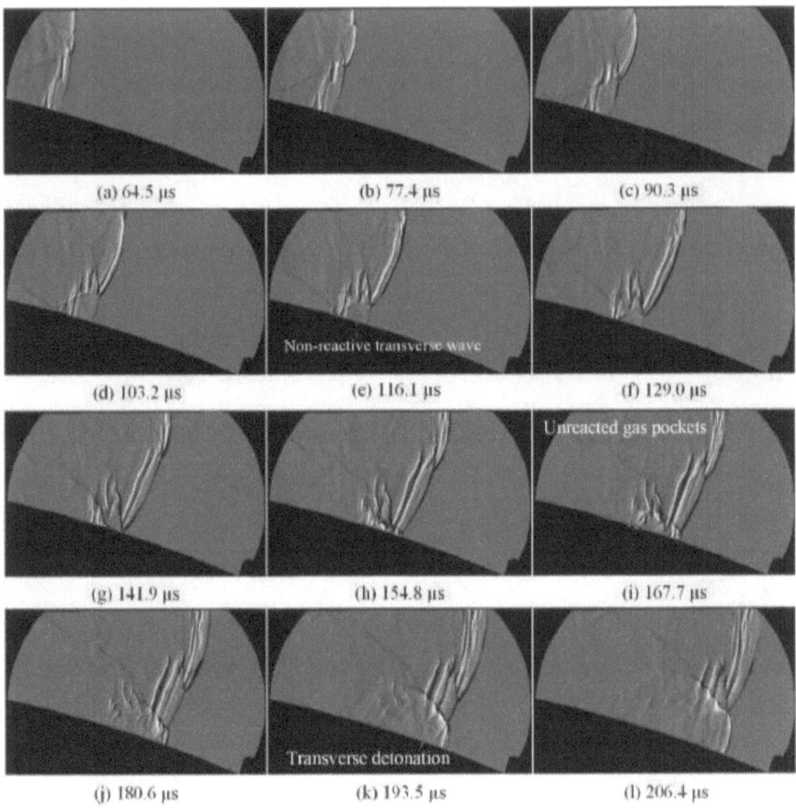



(a) 64.5 μs    (b) 77.4 μs    (c) 90.3 μs

(d) 103.2 μs    (e) 116.1 μs    (f) 129.0 μs

(g) 141.9 μs    (h) 154.8 μs    (i) 167.7 μs

(j) 180.6 μs    (k) 193.5 μs    (l) 206.4 μs

Figure 14: Schlieren photographs illustrating detonation evolution along the large ramp in the mixture of $2H_2/O_2/7Ar$ at the initial pressure of 13.8 kPa.

of both reactive and non-reactive transverse waves existing in marginal detonations [26] and the mechanism of forming unreacted gas pockets due to inert transverse waves [26, 27, 29, 30].

The relationships between the mean propagation speeds and initial pressures are shown in Fig. 15 for detonations in these mixtures, along the two ramps. As the argon dilution increases, detonations of the same initial conditions propagate with a larger velocity deficit, and the critical pressure $p_c$ also increases. This can be interpreted in terms of the gas sensitivity varying with the increase in dilutions of argon. Mixtures with higher argon dilutions are the ones with reduced reaction rates and chemical energy release rates, thus giving rise to larger velocity deficits and limiting pressures. Near the limit, the measured velocity deficits can reach 20% ∼ 25% of the CJ value, as is consistent with that of $2H_2/O_2/2Ar$.

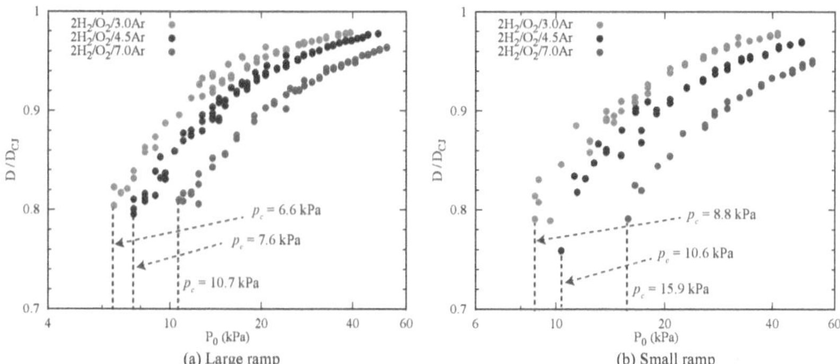

(a) Large ramp         (b) Small ramp

Figure 15: Global mean propagation speeds of (a) along the large ramp and (b) along the small ramp with respect to initial pressures for different mixtures.

## 4. Discussion

### 4.1. Relaxation to quasi-steady state

The investigation of the entrance effects in our experiments shown in Fig. 3 suggest that the relaxation length scale for the wave to adopt a steady state was approximately on the order of the inverse steady state curvature (i.e., the radius of curvature $1/K$ ), i.e., $1/2.17 = 0.46$ m and $1/4.34 = 0.23$ m, respectively for the two ramps investigated.

The relative non-importance of the entrance effects is also borne out from the comparison of the detonation speeds obtained in the first and second half of the channel, shown in Fig. 16 for all the experiments performed. The very good consistency between the mean propagation speeds averaged over the whole section and those over the latter half section further confirms the appropriateness of assuming the macro-scale quasi-steady detonations inside the exponential channels.

In order to investigate whether the conclusion of short transient entrance effects can be generalized, we have further analyzed the numerical results of Radulescu and Borzou [22] for different entrance heights. These new compiled results are shown in Fig. 17. In both simulations, a planar ZND detonation enters a diverging ramp with the constant divergence rate, i.e., $\bar{K} = K\Delta_{1/2} = 0.004$. Two entrance heights are considered, $10\Delta_{1/2}$ and $1\Delta_{1/2}$, where $\Delta_{1/2}$ is the half reaction zone length. Further details can be found in Ref. [22].

The evolution of the detonation front speed recorded along the straight bottom wall, as well as its running time average, are shown in Fig. 17b and Fig. 17c, respectively. From the running time average, it can be observed that it takes a length scale of approximately 200 (in the order of $1/0.004 = 250$) for detonations in both the long and short channels to achieve the quasi-steady state, despite their different entrance heights. In passing, one can also note that minor acceleration effects can be observed near the end of the channels, which is consistent with the slightly smaller curvature found near the end of ramps in Fig. 3 - an indication of the limitation of the quasi-1D assumption in the design of the exponential geometry previously discussed.

Moreover, results in Fig. 17d demonstrate that detonations in both the long and short channels follow the same evolution behaviors, since their running time average speed profiles collapse very well together. It can thus be concluded that the entrance heights have no significant influence on the evolution of detonations. The relaxation length scale appears to correlate well with the radius of curvature dictated by the specific geometry.

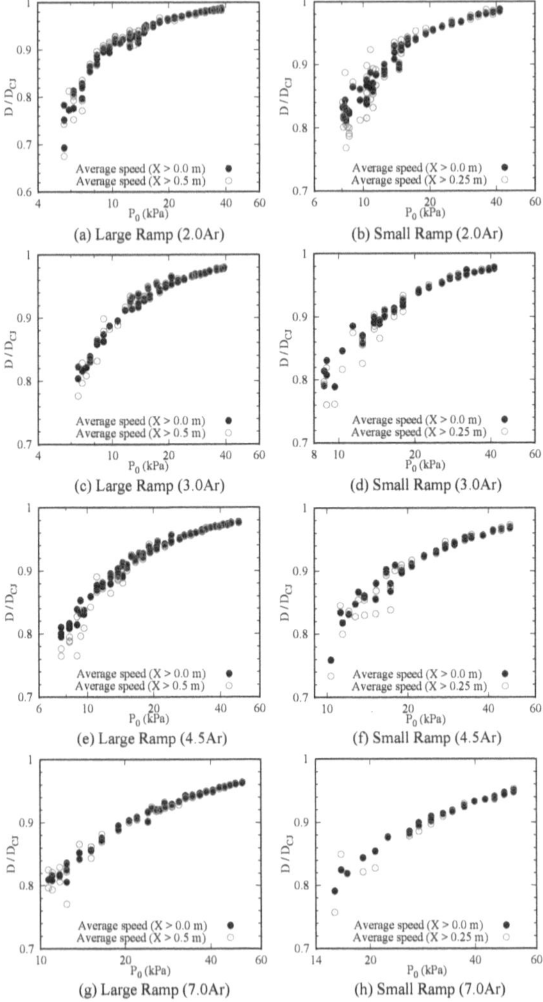

Figure 16: Comparison of the global mean propagation speed ($X > 0.0$ m) with the average speed over the latter half section of the ramp ($X > 0.5$ m for the large one, while $X > 0.25$ m for the small one).

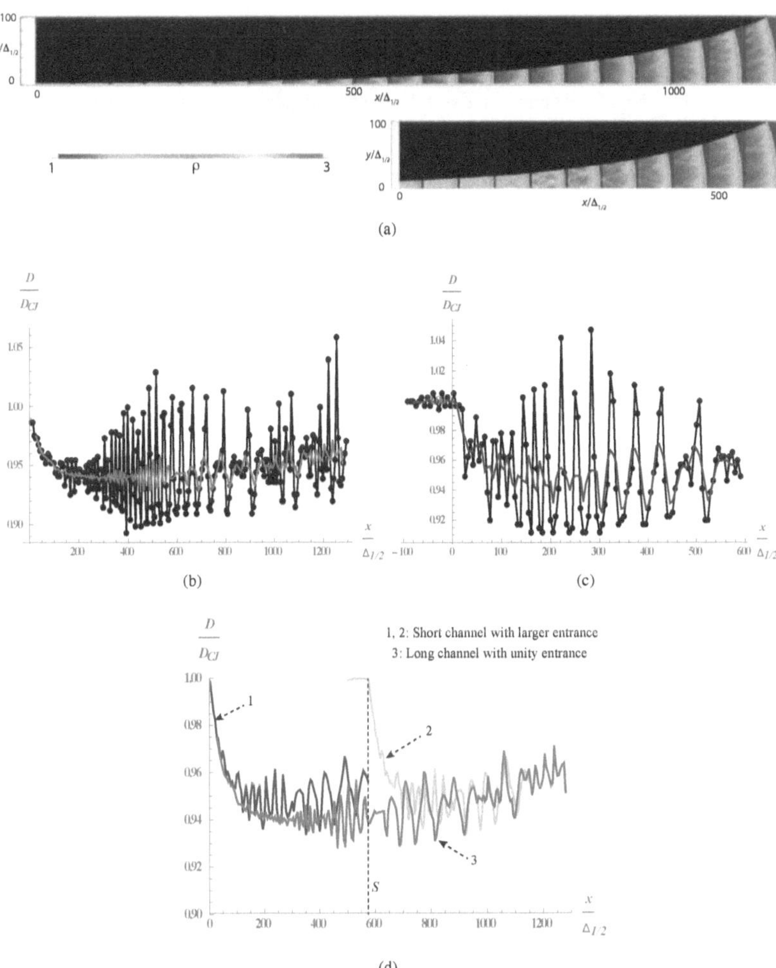

Figure 17: Evolution of a curved detonation in exponential channels with $K\Delta_{1/2} = 0.004$ [22]: (a) the superimposed detonation fronts at different instants for the long channel with unity entrance and the short channel with larger entrance height of $10\Delta_{1/2}$; (b) the detonation speed measured along the straight bottom wall of the long channel and its local time-average; (c) the detonation speed measured along the straight bottom wall of the short channel and its local time-average and (d) comparison of local time-average speeds for the long channel and short one. Profile 2 is obtained by shifting profile 1 to the place $S$, where the long channel and short one have the same channel height.

Note that our finding that the relaxation time and length both scale with $1/K$ is at odds with the quasi-steady state evolution of curved detonations described by the theory of Geometrical Shock Dynamics (GSD) [39, 40], which suggests that the relaxation time to steady state follows a process of diffusion of curvature on the surface of the front controlled by a time scale proportional to $h^2 K$, where $h$ is the channel's thickness. The quasi-steady GSD description requires that the detonation be much thinner than the observation length scales (comparable to the channel height $h$). In our experiments, the detonation enters the curved channel when the channel height is comparable with the reaction zone thickness and comes to a steady state very rapidly. This steady state is then maintained by the constant logarithmic derivative of the channel height.

### 4.2. The generalized ZND model

For the steady, inviscid, reacting quasi-1D flow behind the leading shock front of a detonation wave, the governing equations can be expressed in the frame of reference attached to the shock front as the following system of ordinary differential equations (ODEs) [41]

$$\frac{\mathrm{d}}{\mathrm{d}x'}\left(\rho u A_{tot}\right) = 0 \tag{3a}$$

$$\rho u \frac{\mathrm{d}u}{\mathrm{d}x'} + \frac{\mathrm{d}p}{\mathrm{d}x'} = 0 \tag{3b}$$

$$\frac{\mathrm{d}}{\mathrm{d}x'}\left(h + \frac{1}{2}u^2\right) = 0 \tag{3c}$$

$$u\frac{\mathrm{d}y_i}{\mathrm{d}x'} = \frac{W_i \dot{\omega}_i}{\rho} \quad (i = 1, \cdots, N_s) \tag{3d}$$

where $\rho, u, A_{tot}, p, h, y_i, W_i, \dot{\omega}_i$, and $N_s$ are the mixture density, particle velocity, total cross-sectional area, pressure, enthalpy, species mass fraction, molecular weight, molar production rate of species $i$, and the total number of species. $x'$ is the distance from the shock front under the shock-attached reference. The more convenient form of the above set of equations for computation is

$$\frac{\mathrm{d}p}{\mathrm{d}t} = -\rho u^2 \frac{\dot{\sigma}_{re} - \dot{\sigma}_A}{\eta} \tag{4a}$$

$$\frac{\mathrm{d}\rho}{\mathrm{d}t} = -\rho \frac{\dot{\sigma}_{re} - M^2 \dot{\sigma}_A}{\eta} \tag{4b}$$

$$\frac{\mathrm{d}u}{\mathrm{d}t} = u \frac{\dot{\sigma}_{re} - \dot{\sigma}_A}{\eta} \tag{4c}$$

$$\frac{\mathrm{d}y_i}{\mathrm{d}t} = \frac{W_i \dot{\omega}_i}{\rho} \quad (i = 1, \cdots, N_s) \tag{4d}$$

$$\frac{\mathrm{d}x'}{\mathrm{d}t} = u \tag{4e}$$

with

$$\eta = 1 - M^2, \qquad \dot{\sigma}_{re} = \sum_{i=1}^{N_s}\left(\frac{W}{W_i} - \frac{h_i}{c_p T}\right)\frac{\mathrm{d}y_i}{\mathrm{d}t}, \qquad \dot{\sigma}_A = \frac{u}{A_{tot}}\frac{\mathrm{d}A_{tot}}{\mathrm{d}x'} \tag{4f}$$

where $\dot{\sigma}_{re}$ is the thermicity of ideal gases, while $\dot{\sigma}_A$ is the rate of lateral strain. $M$, $h_i$, and $c_p$

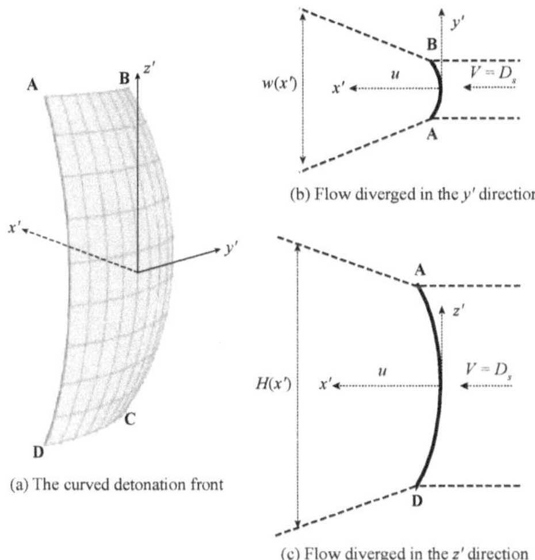



Figure 18: Sketch illustrating the flow diverged behind the curved detonation front under the shock attached reference. Note that the incoming flow propagates along the $x'$ direction.

are the local Mach number, specific enthalpy of species $i$, and mixture specific heat at constant pressure. For the detonation wave, the flow behind the leading shock starts from being subsonic and then accelerates towards the sonic condition due to the positive energy release of $\dot{\sigma}_{re}$. On the other hand, the strain rate of $\dot{\sigma}_A$ by the lateral flow divergence plays the opposite effect of decelerating this flow. As the flow may eventually become supersonic, a singular behavior would appear in Eqs. (4a)-(4c). The local balance of these two competing effects, however, can effectively prevent this singular phenomenon by mathematically letting $\dot{\sigma}_{re} = \dot{\sigma}_A$ happen at the same point for $u = c$ ($M = 1$). As a result, this so-called generalized CJ condition is satisfied with the steady reaction zone solution smoothly passing through the sonic point [41]. With this criterion, the simplified system of governing equations of Eq. (4) can be numerically solved.

The lateral strain rate experienced by the flow in the detonation attached frame can be related to two separate effects in our experiments: the divergence of the flow to boundary layers and that due to an enlarging channel dimension [22]. Consider a curved detonation front $ABCD$ in Fig. 18a, where $x', y', z'$ are the space coordinates under the shock-attached system. The incoming flow (along the $x'$ direction) across the shock diverges in both $y'$ and $z'$ directions, as shown in Figs. 18b and c, respectively. The lateral flow divergence behind the detonation can be related with the shock front curvature through the following expressions as [1, 41]

$$\frac{1}{A_{tot}} \frac{dA_{tot}}{dx'} = K_{eff} \left( \frac{D_s}{u} - 1 \right) \tag{5a}$$

$$\frac{1}{w(x')} \frac{dw(x')}{dx'} = K_{AB} \left( \frac{D_s}{u} - 1 \right) \tag{5b}$$

$$\frac{1}{H(x')} \frac{dH(x')}{dx'} = K_{AD} \left( \frac{D_s}{u} - 1 \right) \tag{5c}$$

where $D_s$ is the detonation speed. $K_{eff}$ is the effective curvature of the curved detonation front $ABCD$, while $K_{AB}$ and $K_{AD}$ are the curvature due to the diverging flow in directions of $y'$ and $z'$, respectively. Since the total cross-sectional area $A_{tot}$ is $A_{tot} = H(x') \times w(x')$, we can thus have

$$\frac{1}{A_{tot}} \frac{dA_{tot}}{dx'} = \frac{d}{dx'}(lnA_{tot}) = \frac{1}{H(x')} \frac{dH(x')}{dx'} + \frac{1}{w(x')} \frac{dw(x')}{dx'} \tag{5d}$$

Therefore, we can further obtain

$$K_{eff} = K_{AD} + K_{AB} \tag{5e}$$

which indicates that for the curved detonation front $ABCD$ in Fig. 18a, its total curvature of $K_{eff}$ includes two parts, one due to the flow diverged in the $y'$ direction while the other one comes from the $z'$ direction. Similarly, for the curved detonation in our exponential channels, its total flow divergence also includes two parts, the one from the geometrical divergence of the exponentially diverging channel in the height direction and the other due to divergence of the flow rendered by the boundary layer growth on the channel side walls. The effective curvature $K_{eff}$, experienced by detonations in the exponential geometry, can thus be expressed as

$$K_{eff} = K + \phi_{BL} \tag{6}$$

where $K$ is the logarithmic area divergence rate of the geometry. For the large ramp, $K = 2.17$ m$^{-1}$, while for the small one, $K = 4.34$ m$^{-1}$. $\phi_{BL}$ represents the contribution of the boundary-layer-induced losses from the channel width direction.

### 4.3. The experimental D(κ) curves

Instead of modelling the boundary-layer-induced lateral flow divergence, Radulescu and Borzou [22] directly evaluated this loss rate of $\phi_{BL}$ from experiments by analytically comparing the experimental data of two ramps with two underlying assumptions: (1) detonations propagating inside the exponentially diverging channel of different expansion ratios have the same constant $\phi_{BL}$ since the channel's dimension of the width is unchanged; (2) for the same mixture, it has a unique relation between the velocity deficits and the losses. As a result, the effective curvature $K_{eff}$ of the global front can be calibrated by collapsing together the experimental $D(\kappa)$ curves of detonations in the large and small ramp experiments, and then the loss rate $\phi_{BL}$ due to boundary layers can be derived [22]. Figure 19 shows the experimentally obtained $D(\kappa)$ curves, characterizing the relationships between the detonation velocity (normalized by the ideal CJ speed) and the lateral flow divergence, for all the mixtures involved in this study. Note that the abscissa is the non-dimensional loss obtained by multiplying the curvature with the ZND induction zone

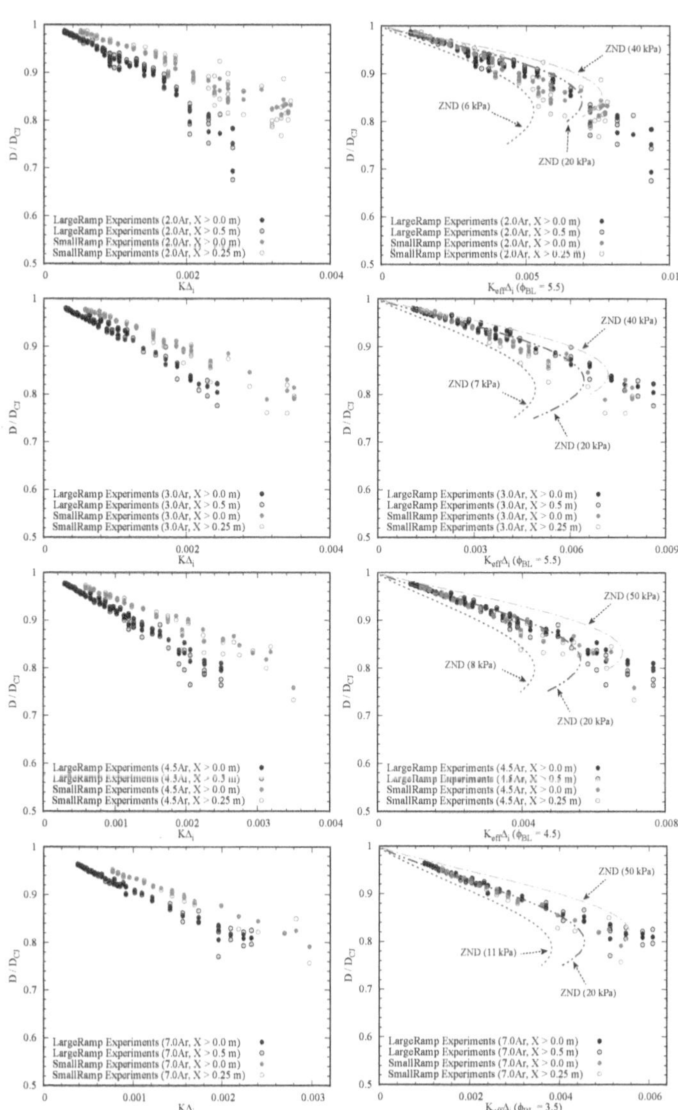

Figure 19: The non-dimensional $D - K$ characteristic relationships obtained experimentally and predicted from the generalized ZND model using the detailed San Diego chemical mechanism.

length $\Delta_i$, which is taken as the distance between the CJ detonation front and the location of its peak thermicity. It was calculated by using the Shock and Detonation Toolbox (SDToolbox) [42] under Cantera's framework [43] with the San Diego chemical reaction mechanism (Williams) [44]. This normalization method is the same as that adopted by Radulescu and Borzou [22]. Moreover, theoretical works have demonstrated that, for a specific mixture, there exists a unique relationship between the curved detonation velocity deficit and the curvature normalized by the induction zone length [41, 45, 46]. In Fig. 19, the graph in the right column is the collapsed $D/D_{CJ} - K_{eff}\Delta_i$ correlation, after calibrating the effective curvature $K_{eff}$ from the $D/D_{CJ} - K\Delta_i$ curve in the corresponding left graph, by including the boundary layer effects. In the work of Radulescu and Borzou [22], the same unique value of $\phi_{BL} = 5.5$ m$^{-1}$ was found to permit collapsing all data of two ramps in both mixtures of $2C_2H_2/5O_2/21Ar$ and $C_3H_8/5O_2$. However, the present study found that $\phi_{BL}$ can be varied for different mixtures, with 5.5 m$^{-1}$ for $2H_2/O_2/2.0Ar$ and $2H_2/O_2/3.0Ar$, while 4.5 m$^{-1}$ and 3.5 m$^{-1}$ for $2H_2/O_2/4.5Ar$ and $2H_2/O_2/7.0Ar$, respectively.

In the recent works of Kudo et al. [47] and Nakayama et al. [48, 49] of curved gaseous detonations in rectangular-cross-section curved channels, they also obtained the meaningful characteristic $D(k)$ curves, which are the first extension of the $D(\kappa)$ theory experimentally from condensed-phase detonations to gas-phase detonations. When the detonation cell size was small enough, i.e., in the order of 0.1~1.0 mm, they were able to obtain the curved detonations in quasi-steady conditions. An interesting question that arises is why the boundary layer effects can be negligible in their works while appear to be significant in the present work, since their channel depth of 16 mm is in the same order with that of the current exponential channels. This can be first clarified by the fact that the cell sizes of detonations in the present study are comparable to, if not larger than, the channel width, while in works of Nakayama et al., the cell size is much smaller. As a result, detonations of much longer characteristic reaction zones in the present experiments experienced more significant boundary-layer-induced losses. The other explanation can be interpreted with the detonation loss induced by the geometry. The exponential-geometry-induced mean front curvature in the present work has been found to be comparable to that due to boundary layers. However, in works of Nakayama et al., the local curvature induced by the geometry ranges from 20 ~ 200 m$^{-1}$ (evaluated from Fig. 7 of Ref. [48]), which is much larger than the boundary-layer-induced loss rate estimated from current experiments. Therefore, as a result of the difference between geometries and initial conditions, the boundary layer can play the role of varied significance.

### 4.4. Comparisons with the generalized ZND model

#### 4.4.1. The ZND model predicted $D/D_{CJ} - K_{eff}\Delta_i$ curves

To start with, the theoretically predicted relationship between the effective curvature $K_{eff}$ and the detonation speed has been obtained by solving the ODE system of Eq. (4), with the developed custom Python code [22] working under the framework of SDToolbox and Cantera. The San Diego reaction mechanism was applied for describing the realistic chemical kinetics. These calculations were conducted at three initial pressures covering the experimental range. Normalizing the varied divergence rates $K_{eff}$ by the ZND induction zone length $\Delta_i$ at these pressures permits getting the theoretical $D/D_{CJ} - K_{eff}\Delta_i$ curves, which have been shown in Fig. 19. These comparisons show that the experiments are generally in very good agreement with the extended ZND model predictions for small and moderate lateral strain, except near the limit. Detonations from experiments were able to propagate beyond the maximum lateral strain predicted by the steady ZND model, with larger velocity deficits, thus demonstrating the role

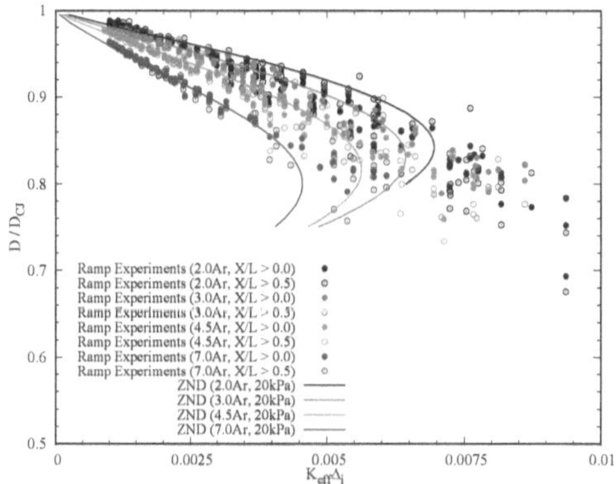

Figure 20: The $D/D_{CJ} - K_{eff}\Delta_i$ characteristic curves obtained from experiments and predicted from the quasi-1D ZND model for different mixtures.

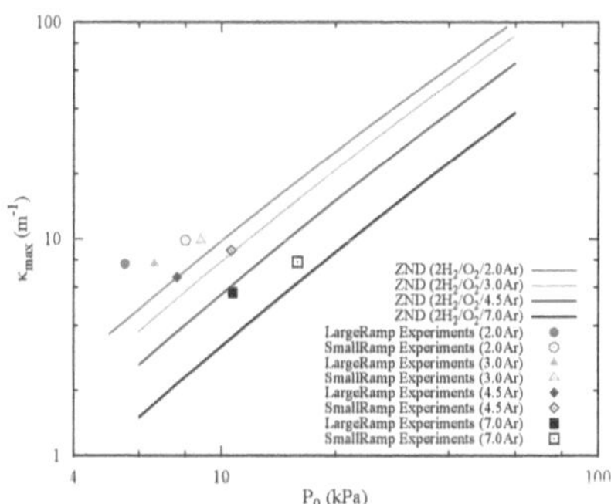

Figure 21: Critical curvature as a function of initial pressures obtained from experiments and predicted from the quasi-1D ZND model for different mixtures.

of cellular structures in enhancing the detonability of gaseous detonations [50]. The $D/D_{CJ} - K_{eff}\Delta_i$ characteristic relationships of both the experiments and ZND model predictions for all the mixtures, as summarized in Fig. 20, demonstrate that detonations with less argon dilutions can propagate with more losses. It is thus indicative of the effects of argon dilutions in reducing the detonability of $H_2/O_2/Ar$ detonations, consistent with the computations by Klein et al. [41].

### 4.4.2. The ZND model predicted critical curvature $\kappa_{max}$

The effects of argon dilutions in reducing the detonation detonability are more evident in Fig. 21, which illustrates the relationship between the initial pressures and the ZND model predicted critical curvature $\kappa_{max}$, above which detonations are not possible at a certain initial pressure. With the reduction of the mixture sensitivity by lowering the initial pressure, the critical curvature also decreases with the reduced detonability. Moreover, the results in Fig. 21 further demonstrate that detonations in experiments can propagate with a larger critical curvature than that predicted by the generalized ZND model, suggesting the higher detonability of real detonations in experiments.

### 4.4.3. The ZND model predicted detonation speed as a function of initial pressures

Finally, the theoretically predicted detonation speed as a function of initial pressures has also been calculated, as shown in Fig. 22 illustrating the comparisons of the experiments with the predictions made with the real chemistry. Such ZND model predicted relationships were obtained through three approaches for evaluating the boundary-layer-induced flow divergence, either directly from experiment, or by modeling the negative displacement thickness in the case of turbulent or laminar boundary layers.

(1) The first method assumed the constant boundary-layer-induced loss rate $\phi_{BL}$, which has been directly calibrated from experiments. The lateral strain rate $\dot{\sigma}_A$ can thus be expressed as

$$\dot{\sigma}_A = (D_s - u)(K + \phi_{BL}) \tag{7}$$

(2) According to Fay's theory on boundary layer mechanism, its effect can be modelled by inviscid flow in a streamtube with a negative displacement of the boundary layer, as a result of the boundary layer acting as a mass sink by removing the mass from the core flow [4]. For the present experimental configuration, since the channel height is much larger than the channel width, the boundary layer effects on the top and bottom curved wall can be reasonably neglected. The effective width of the diverged flow behind the detonation is thus given by $w(x') = w + 2\delta^*(x')$, where $\delta^*(x')$ is the boundary layer negative displacement thickness . As a result, the rate of flow divergence due to the boundary layer growth on side walls can be evaluated as

$$\frac{1}{w(x')}\frac{dw(x')}{dx'} = \frac{2}{(w + 2\delta^*)}\frac{d\delta^*}{dx'} \tag{8}$$

In Fay's work, he adopted Gooderum's empirical turbulent boundary layer thickness relation [51] as the displacement thickness behind the detonation wave. The relation is given by

$$\delta^*(x') = 0.22\left(x'\right)^{0.8}\left(\frac{\mu_e}{\rho_0 D_s}\right)^{0.2} \tag{9}$$

where $\mu_e$ and $\rho_0$ are the post-shock state viscosity and the initial density, respectively. This turbulent boundary layer displacement thickness relation has then been applied in a large number of subsequent works investigating detonation velocity deficits [5, 6, 9, 11–15]. The boundary-layer-induced flow divergence rate is thus a function of the distance behind the leading shock. Consequently, using this turbulent boundary layer displacement thickness relation, the lateral strain rate $\dot{\sigma}_A$ is

$$\dot{\sigma}_A = (D_s - u)K + u\frac{0.352}{w + 2\delta^*}\left(\frac{\mu_e}{\rho_0 D_s}\right)^{0.2}\left(\frac{1}{x'}\right)^{0.2} \tag{10}$$

(3) For laminar boundary layers, the displacement thickness was obtained by solving Mirels'

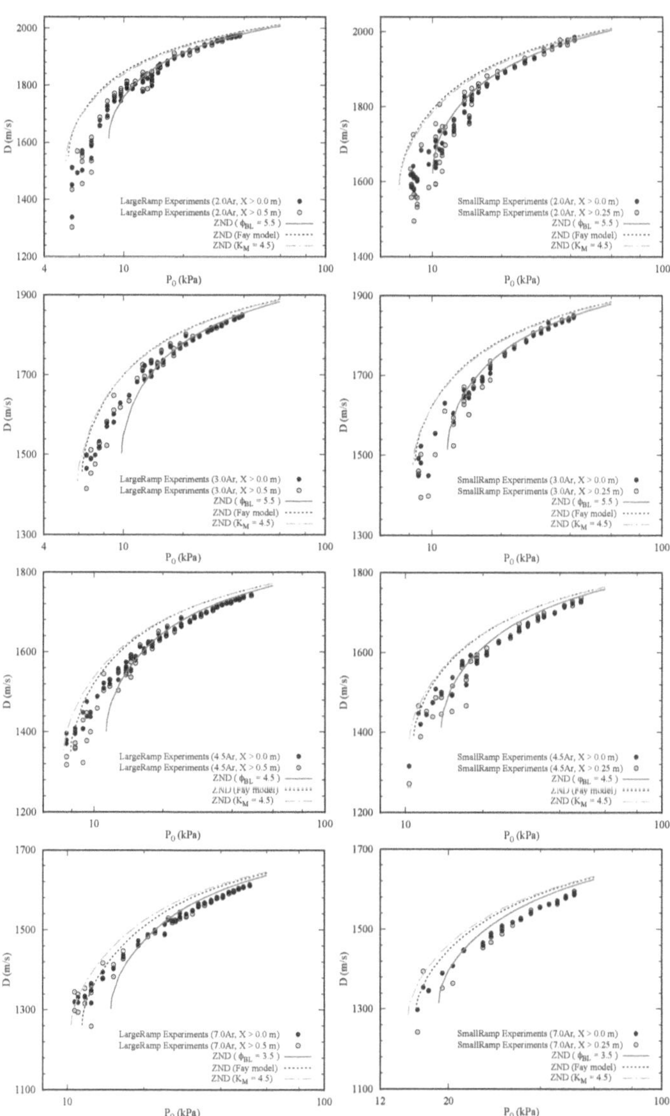



Figure 22: Comparisons of the experimentally obtained average speeds with the generalized ZND model predictions for various mixtures; the red line adopts the experimentally obtained constant divergence rate of $\phi_{BL}$ due to the boundary layer for the ZND model prediction, while the broken blue line and green line are the predictions made with the turbulent and laminar boundary layer assumptions, respectively.

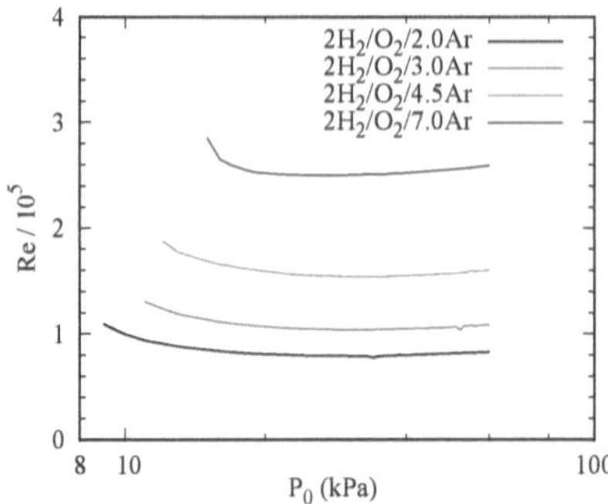

Figure 23: Reynolds number $Re = (\rho_s u_s/\mu_s) x_H$ for $H_2/O_2/Ar$ detonations with the same lateral flow divergence rate as those in large ramp experiments under varied initial pressures.

compressible laminar boundary layer equations [52] (see the Appendix)

$$\delta^*(x') = 4.5 \sqrt{\frac{\mu_e x'}{\rho_e u_e}} \tag{11}$$

where we have $\rho_e u_e = \rho_0 D_s$ via the mass conservation. The lateral strain rate $\dot{\sigma}_A$ then is

$$\dot{\sigma}_A = (D_s - u)K + u \left(\frac{4.5}{w + 2\delta^*}\right)\left(\frac{\mu_e}{\rho_0 D_s}\right)^{0.5}\left(\frac{1}{x'}\right)^{0.5} \tag{12}$$

The comparisons in Fig. 22 show that the ZND model predictions, obtained with Fay's turbulent boundary layer displacement thickness relation and Mirels' laminar boundary layer solutions, both have the very close results. They can relatively well predict the experiments. And surprisingly, they also appear to capture the near-limit detonation dynamics quite well, including both the limit pressures and velocity deficits. Moreover, we also estimated the Reynolds number $Re = (\rho_s u_s/\mu_s) x_H$, as shown in Fig. 23, for $H_2/O_2/Ar$ detonations in large ramp experiments. Note that $\rho_s$, $u_s$, and $\mu_s$ are the post-shock parameters, i.e., density, particle velocity (in the lab frame reference), and viscosity, while $x_H$ is taken as the characteristic hydrodynamic thickness between the leading shock and the CJ sonic surface. The results from Fig. 23 indicate that the boundary layer behind $H_2/O_2/Ar$ detonations is probably laminar, since the Reynolds number is smaller than the critical one of $Re_c \approx 0.5 \times 10^6 \sim 4.0 \times 10^6$ for the shock-induced boundary-layer transition from laminar to be turbulent [51, 53–55]. This finding concurs with the previous conclusion of Liu and Glass [56] and Damazo et al. [57] that the boundary layer behind stoichiometric hydrogen/oxygen detonations is laminar. Therefore, Fay's turbulent boundary layer displacement thickness relation for evaluating the boundary-layer-induced flow divergence is questionable. It is not clear at present why this turbulent boundary layer assumption somehow can work well for predicting the experiments.

On the other hand, it can also be clearly seen from Fig. 22 that the experiments are in excellent agreement with the predictions, made with the constant boundary-layer-induced loss rate of $\phi_{BL}$

directly evaluated from experiments, for detonations with small and moderate velocity deficits well above the limit. While approaching the limit, these predictions deteriorate. Real detonations can propagate at lower initial pressures, where the steady ZND model predicts failure. The first reason for this discrepancy near the limit can be attributed to the possible promotion mechanism of strong reactive transverse waves, especially the transverse detonations, which help extend the propagation limits to lower pressures than that predicted by the generalized ZND model neglecting cellular structures. Secondly, the assumption of a constant global curvature for the leading shock front, well above the limit, is not applicable to detonations near the limit, as can be concluded from observations of the structures of detonation fronts at relatively low pressures, e.g., see Fig. 4. It thus results in failure of the theoretical predictions for these near-limit detonations from experiments.

The observed differences between the ZND model prediction and experiments observed near the limits may also be affected by the enhanced instability of attenuated detonations with lower shock temperatures and longer ignition delays compared to the reaction time scales. Radulescu [19] introduced the parameter $\chi$ for characterizing the stability of detonations under different thermodynamic conditions. Detonations in mixtures of higher $\chi$ are more unstable to perturbations in the reaction zones than those with lower $\chi$. The mathematical expression of $\chi$ is

$$\chi = \left(\frac{E_a}{RT_s}\right)\left(\frac{t_{ig}}{t_{re}}\right) \tag{13}$$

where $E_a/RT_s$ is the usual non-dimensional activation energy, $T_s$ is the temperature behind the leading shock, $R$ is the specific gas constant, and $t_{ig}/t_{re}$ is the ratio of ignition to reaction time. Figure 24 shows, for all the mixtures, the relationships of these parameters as a function of the leading shock velocity, normalized with the ideal CJ speed. Of noteworthy is that each solid circle represents one experiment, and the shock speed corresponds to the experimentally measured mean propagation velocity over the whole ramps. One can observe that decreasing the detonation front speed by increasing the velocity deficits results in the reduced post-shock temperature, increased activation energy, and larger ratios of the ignition time relative to the reaction time, thereby enhancing the sensitivity of the reaction rates to relaxations in the reaction zones. As such, detonations tend to become more unstable with increased $\chi$. The considerable increment of $\chi$, however, occurs when the velocity deficit is relatively large, especially near the limit, as can be observed from Fig. 24d. It can thus be speculated that the significantly increased instability, as a result of considerably large velocity deficits, leads to the incapability of the extended ZND model to predict detonation dynamics at relatively low initial pressures near the limit.

### 4.5. Why can the steady 1D ZND model predict the $H_2/O_2/Ar$ cellular detonation dynamics?

The above analysis has quantitatively showed that the generalized ZND model with lateral strain rate can predict very well the experiments of $H_2/O_2/Ar$ cellular detonations, except some departures for the near-limit detonation dynamics. The question that arises is what results in such excellent predictability, given the detonation structure is always cellular.

A tentative answer comes from the analysis of the relevant length and time scales in the detonation cellular structure. For hydrogen-oxygen-argon detonations, the induction time scales and length scales are much shorter than the global reaction zone, as can be seen from the ZND calculations of Fig. 25. 90% of energy release occurs at a distance an order of magnitude longer than the induction zone thickness. Since the reaction zone is not thermally sensitive, spatial

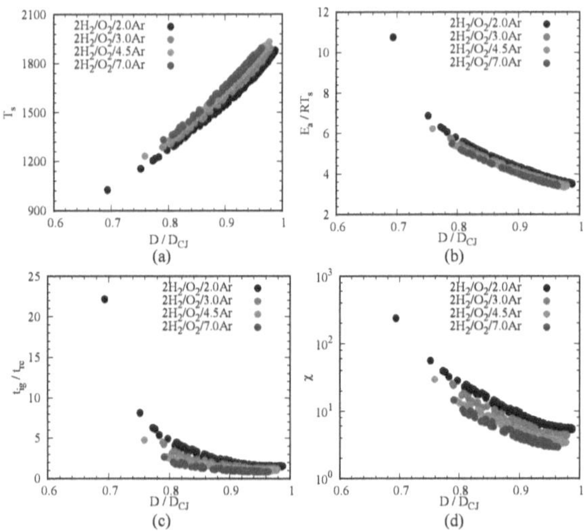

Figure 24: Parameters of (a) post-shock temperature $T_s$, (b) reduced activation energy $\theta = E_a/RT_s$, (c) the ratio of ignition time to reaction time, and (d) the stability parameter $\chi$ as a function of the shock speed normalized with the CJ velocity. The ignition time here is defined by the time to peak thermicity, while the reaction time is defined as the inverse of the maximum thermicity, as proposed by Radulescu [19]. They were calculated with the constant volume explosion at the mean propagation speed with the real chemistry.

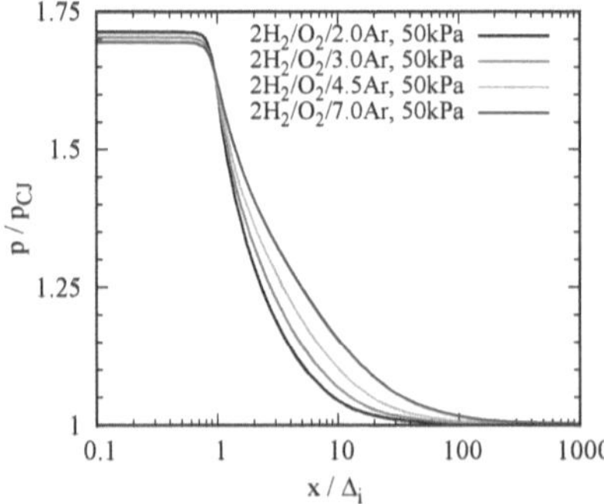

Figure 25: The ZND structure of CJ detonations: pressure profile as a function of the distance normalized by the induction zone length.

variations in the induction times associated with the cellular structure do not appreciably change the reaction zone structure. Since the dynamics are controlled by the acoustic time over the whole reaction zone structure, these are expected to be insensitive to the induction zone variations, which are much faster. For hydrocarbon detonations studied by Radulescu and Borzou [22], the opposite is true and the reaction zone thicknesses are much shorter as compared to the induction zone thickness. For these mixtures, the ZND model fails to capture the dynamics.

## 5. Conclusion

In the present study, experiments of very regular cellular detonations, propagating inside the exponentially diverging channels with two different constant area divergence rates, were performed in mixtures of stoichiometric $H_2/O_2$ of varied argon dilutions. The propagation characteristics were demonstrated and analyzed in detail. The results showed that detonations, well above the limit, were uniformly curved with small-sized cellular structures of constant cell sizes, and can be reasonably assumed to propagate in quasi-steady state at the macro-scale with a constant mean front curvature. The experimentally measured cell sizes showed a strong dependence on the detonation velocity deficits. When the mean propagation speed dropped to about $0.8D_{CJ}$, detonations started to propagate in the mechanism of single-headed detonations, organized with a distinctive transverse detonation, with the detonation cell size larger than the ideal one in the order of $10 \sim 15$. The stabilization mechanism of single-headed detonations with saliently enlarging cells suggests a higher growth rate of the curved detonation front area than that of the intrinsic transverse instability. The Go/No-Go phenomenon has also been demonstrated for detonations near the very limit, due to its stochastic property. Also, both the reactive and non-reactive transverse waves were observed for detonations propagating at low pressures near the limit.

The experimental $D/D_{CJ} - K_{eff}\Delta_i$ curves were obtained by collapsing the $D/D_{CJ} - K\Delta_i$ relationships of both the large ramp and small ramp. As a result, the equivalent loss rate $\phi_{BL}$ due to boundary layers were directly derived. It was found that detonations of lower argon dilutions can propagate with more losses, indicating the effects of argon addition in decreasing the detonability of $H_2/O_2/Ar$ detonations. The same conclusion can also be drawn from the decrease of critical curvature with the increasing argon dilutions. The variation of velocity deficits with the flow divergence was found in very good agreement with the predictions made with steady ZND model with lateral strain rate, for small and moderate divergence and velocity deficits. Nevertheless, the model under-predicted the limiting lateral strain and velocity deficits for detonations near the limit, thus illustrating the effects of cellular structures in enhancing the detonability.

Furthermore, comparisons between the experimentally obtained average speeds and the generalized ZND model predictions were performed in terms of different initial pressures. The results showed that predictions made with the experimentally obtained constant loss rate $\phi_{BL}$ can excellently predict the experiments, except some departures for the near-limit detonations. On the other hand, the predictions made with Fay's turbulent boundary layer displacement thickness relation and Mirels' laminar boundary layer solutions both can relatively well predict the experiments. However, the estimated Reynolds number casts doubt on the turbulent boundary layer assumption, while appears to support the finding that the boundary layer behind hydrogen/oxygen detonations is laminar. Finally, the promotion effects by the strong reactive transverse waves or transverse detonations, the limitation of the quasi-1D assumption near the limit, and the considerably increased instability were proposed for clarifying the discrepancies between the theoretical predictions and experiments for the near-limit detonations. The much shorter induction zone length and time scales as compared to the global reaction zone were proposed for clarifying the predictability of the real $H_2/O_2/Ar$ cellular detonation dynamics by the extended 1D ZND model with lateral strain rate.

## 2.3 Role of Instability on the Limits of Laterally Strained Detonation Waves

**Status:** The preliminary work of this paper was presented at the 27th International Colloquium on the Dynamics of Explosions and Reactive Systems (ICDERS) in Beijing, China in 2019 [76]. It has been published in *Combustion and Flame.*

The present work examines the role of instability and diffusive phenomena in controlling the limits of laterally strained detonation waves. It further extends the exponential horn technique of Radulescu and Borzou [17] to unstable detonations in three other hydrocarbon-oxygen mixtures with varied levels of cellular instability, i.e., $CH_4/2O_2$, $C_2H_4/3O_2$, and $C_2H_6/3.5O_2$. Firstly, the characteristic $D - K$ relationships for these detonations were obtained from experiments, and quantitatively compared with the theoretical predictions. The link of departure between experiments and the generalized ZND model predictions has been established. Employing the experimental $D - K$ data, empirical global one-step reaction models were then developed for the hydrodynamic macroscopic average description of cellular detonation dynamics. Finally, the intensified auto-ignition mechanism assisted by the turbulent diffusive burning of the unreacted gas pockets was found to enhance the global rates of energy release, as compared to the laminar ZND wave, thereby resulting in the significantly shortened reaction zone lengths observed in experiments.

# Role of Instability on the Limits of Laterally Strained Detonation Waves

## 1. Introduction

Very recently, Xiao and Radulescu [1] performed experiments with hydrogen-oxygen-argon ($H_2/O_2/Ar$) cellular detonations in exponentially diverging channels. They found that, despite the cellular structures, the detonation dynamics can be excellently predicted from the generalized Zeldovich-von Neumann-Doering (ZND) model with lateral strain rates [2] using detailed chemical kinetics. However, this excellent agreement departs from the earlier experiments of Radulescu and Borzou [3] in more unstable detonations, where the experiments disagreed with the theoretical predictions. The question thus arises whether there is the link of departure between experiments and the steady 1D ZND model predictions to detonation instability. The present work addresses this question.

In practice, detonations usually propagate in the presence of lateral strain [4, 5]. For example, in either weakly confined (e.g., see Refs. [6–11]) or varying-cross-section (e.g., see Refs. [1, 3, 12]) or curved geometries (e.g., see Refs. [13–17]), detonations tend to be curved with the flow experiencing lateral divergence in the reaction zone. A typical case can be seen in rotating detonation engines (RDEs) [18–21], where detonations are weakly confined by the burned products of the previous rotation cycle and the flow behind the front locally experiences side expansions, resulting in lateral strain. The other important scenario is related with detonations in small-sized confined geometries, such as narrow channels or tubes, where detonations are subject to significant losses induced by boundary layers [22–39]. In the shock-attached frame of reference, the boundary layer acts as a mass sink from the inviscid core flow [24] and results in the flow diverging in reaction zones [40, 41], thereby giving rise to lateral strain. This boundary-layer-induced lateral flow divergence can thus be modelled by the negative displacement of boundary layers [24]. Lateral strain has been generally known to significantly impact the detonation dynamics, i.e., decreasing the propagation speeds lower than the theoretical Chapman-Jouguet (CJ) velocities [24, 26, 27, 32, 34–39, 42], increasing the limit pressures [26, 32, 34, 36] as well as cell sizes [7, 11, 37, 42–44].

Extensive efforts have been made in quantifying the response of detonation dynamics to lateral strain for detonations in narrow channels or small tubes (e.g., see Refs. [24, 26, 27, 32, 34–39]). The experimentally obtained detonation mean propagation speed dependence on "lateral flow divergence" was then compared to the extended ZND models [2, 24, 26, 27]. Note that in these works, the "lateral flow divergence" was interpreted in terms of the characteristic geometry length scale (e.g., the channel depth or the tube diameter) normalized by the length scale of the kinetics, such as the detonation cell size [43] or the ZND detonation induction zone length [2, 32]. In general, all the comparisons showed that the steady 1D ZND model can relatively well predict the dynamics of weakly unstable detonations with regular cellular structures, while much worse for the highly unstable detonations with more irregular cellular structures. Specifically in terms of detonation limits, as compared to the steady ZND counterpart, unstable detonations in experiments can propagate at substantially larger limiting flow divergence with much larger velocity deficits [32, 33, 45]. Despite these findings, whether there is a correlation between the departure and detonation instability is still not clear. Moreover, no effective strategies have been proposed in improving the theoretical predictions of the unstable detonation dynamics.

In an attempt to tackle these questions, Radulescu and Borzou [3] recently formulated a novel experimental technique involving a pair of exponentially diverging ramps. The key feature of the ramp is keeping the logarithmic area divergence rate as a constant, which can enable the macro-scale detonations to propagate in quasi-steady conditions with a constant mean lateral strain rate [1, 3]. As a result, the unique characteristic relationships between the detonation mean propagation speed and its lateral flow divergence can be precisely obtained in an unambiguous manner, thus permitting meaningful comparisons between experiments and the steady ZND model predictions. In their works, two mixtures of different detonation instability were tested, i.e., the weakly unstable argon-diluted acetylene-oxygen ($2C_2H_2/5O_2/21Ar$) and highly unstable propane-oxygen ($C_3H_8/5O_2$) detonations. Again, they found a much poorer predictability of the unstable $C_3H_8/5O_2$ detonation dynamics than that of the more stable $2C_2H_2/5O_2/21Ar$ detonations. Besides these comparisons, they extracted the empirical reaction models from the experimental data for the global average description of the detonation dynamics. The potential usefulness of these calibrated global rate laws has been demonstrated by Radulescu [46] in predicting the initiation and failure phenomena in relevant $C_3H_8/5O_2$ detonation experiments [47]. On the other hand, for highly unstable $C_3H_8/5O_2$ detonations, the calibrated activation energy was found to be significantly lower and the reaction kinetics faster in comparison to the ZND model with the underlying chemistry. This fact highlights the role of diffusive processes, involved in the burn-out of unreacted gas pockets, in suppressing the thermal character of the ignition mechanism [48, 49]. Such diffusive phenomenon is essential in resolving the "detonation paradox" that inviscid detonation simulations predict the incorrect trend for the role of instability in real gaseous cellular detonations [50].

Therefore, by further extending the well-posed exponential ramp technique, the present study aims to examine how the instability and diffusive processes affect detonation limits subject to lateral strain in a range of mixtures with varying levels of cellular instability, i.e., stoichiometric methane-oxygen ($CH_4/2O_2$), ethylene-oxygen ($C_2H_4/3O_2$), and ethane-oxygen ($C_2H_6/3.5O_2$). Particularly, it is of interest to establish whether there is the link of departure between experiments and the generalized ZND model predictions to detonation instability. Finally, useful empirical global reaction models are presented for capturing the detonation dynamics.

## 2. Experimental Details

Experiments were conducted in a 3.4 m long aluminium rectangular channel with an internal height and width of 203 mm and 19 mm, respectively. A sketch of the experimental set-up is shown in Fig. 1, which is the same as that adopted by Xiao and Radulescu [1] and Radulescu and Borzou [3]. The shock tube consists of three parts, a detonation initiation section, a propagation section, and a test section. In the initiation part, the mixture was ignited by a high voltage igniter (HVI), which could store up to 1000 J with the deposition time of $2\,\mu s$. Mesh wires were inserted for facilitating the detonation formation. Three 113B24 and five 113B27 piezoelectric PCB pressure sensors ($p_1 \sim p_8$), as shown in Fig. 1, were mounted flush on the top wall of the shock tube for recording pressure signals.

In the test section, two different polyvinyl-chloride (PVC) ramps were used in our experiments. The key characteristic of the ramp is the property of retaining a constant logarithmic area divergence rate ($K = \frac{d(lnA(x))}{dx}$) of the cross-sectional area $A(x)$. This important feature enables detonations at the macro-scale to propagate in quasi-steady conditions [1, 3]. For the large ramp, the logarithmic area divergence rate was $2.17\ \mathrm{m}^{-1}$, while for the small one, the rate was $4.34\ \mathrm{m}^{-1}$. At the entrance, a protruded rounded tip was kept for minimizing the effects of shock reflection on the detonation front. The initial gap between the ramp tip and the top wall of the channel was 23 mm in height. The height between the exponentially curved wall and the top wall is given by $y_{wall} = y_0 e^{Kx}$ for $x > 0$ and by $y_{wall} = y_0 = 23$ mm otherwise.

The test mixtures were stoichiometric methane-oxygen ($CH_4/2O_2$), ethylene-oxygen ($C_2H_4/3O_2$), and ethane-oxygen ($C_2H_6/3.5O_2$). Each of them was prepared in a separate mixing tank by the

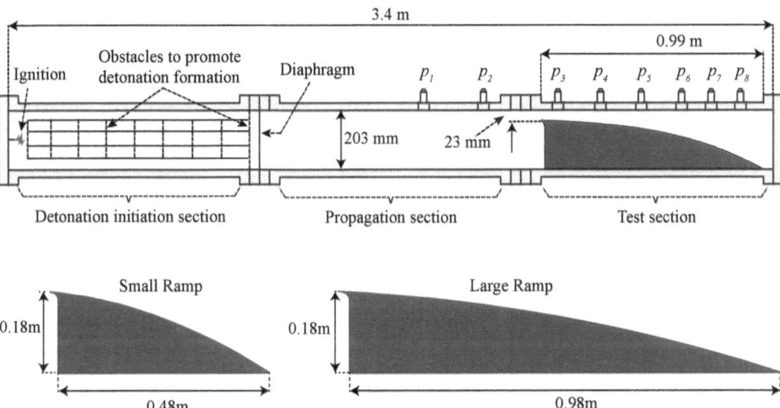



Figure 1: Experimental set-up for diverging detonation experiments.

method of partial pressures and was then left to mix for more than 24 hours. Before filling with the test mixture in every single experiment, the shock tube was evacuated below the absolute pressure of 90 Pa. The mixture was then introduced into the shock tube through both ends of the tube at the desired initial pressure with an accuracy of 70 Pa. Both the schlieren [51] and large-scale shadowgraph [52] techniques were alternatively adopted for visualizing the detonation evolution process along the exponential ramp. The Z-type schlieren was set up with a vertical knife edge using a light source of 360 W. The resolution of the high-speed camera was $384{\times}288$ px$^2$ with the framing rate of 77481 fps (about 12.9 $\mu$s for each interval). For the shadowgraph photos, the resolution of the high-speed camera was $1152{\times}256$ px$^2$ with the framing rate of 42049 fps (about 23.78 $\mu$s for each interval). The light source for the shadowgraph system was provided by a 1600 W Xenon arc lamp. The exposure time for the schlieren and shadowgraph setup was 0.44 $\mu$s and 1.0 $\mu$s, respectively.

## 3. Experimental Results

### 3.1. Detonation propagation characteristics

#### 3.1.1. Methane-oxygen detonations

Figure 2 shows the reaction zone structures of methane-oxygen ($CH_4/2O_2$) detonations along the large ramp at initial pressures relatively well above the propagation limit of $p_c \approx 12.0$ kPa, below which detonations fail to self-sustain. The detonation propagated from left towards right. As a result of the geometrical area divergence, detonation fronts are curved. The other main features are the long-tailed transverse shocks, the fine-scale dense unreacted gas pockets breaking away from the very rough front, and also the trailing scattered pressure pulses behind the spotty frontal structure. These characteristics have been previously reported by Radulescu et al. [48, 53] and Kiyanda and Higgins [54] for methane-oxygen detonations in straight channels. One can have a clearer observation of the formation process of the unreacted gas pockets or islands in the propagation of much lower initial pressure detonations, as shown in Fig. 3. It illustrates

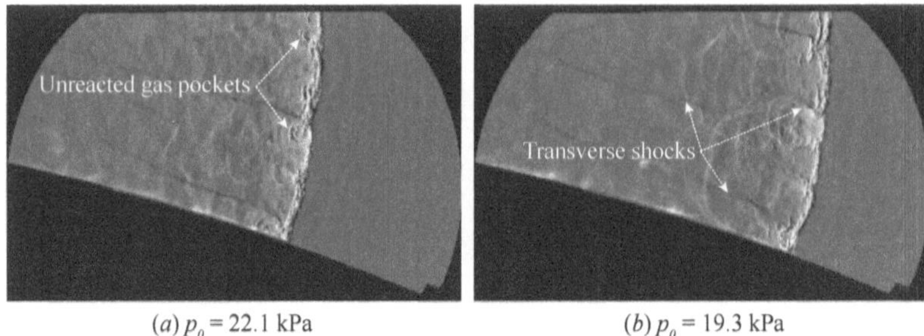



(a) $p_0$ = 22.1 kPa          (b) $p_0$ = 19.3 kPa

Figure 2: Structures of $CH_4/2O_2$ detonations propagating along the large ramp at relatively high initial pressures.

the detailed interactions of triple points in the three-mode (i.e., one cell and half) detonation evolution process along the large ramp. Due to the longer ignition delays for the shocked gases processed by the weaker incident shock, the observed reaction zones are thicker than that behind the stronger Mach shock. The tongue-like region of unreacted gases accumulated along the shear layers increases until the collision of triple points, giving rise to the dense unburned gas island peeled away from the main front, as can be seen from Fig. 3c to Fig. 3f. In the experiment, this unreacted gas pocket was consumed within five frames, i.e., approximately 65 $\mu$s, from Fig. 3e to Fig. 3j. Noteworthy is that this burn-out time of unreacted pockets is approximately the timescale of the detonation cell, consistent with the previous observations [48, 49, 53, 54].

Such behavior can be further confirmed in the single-mode (or single-head, i.e., half cell) detonation propagation, as shown at some length in Fig. 4. The prominent feature of the detonation in Fig. 4a is the highly turbulent tongue-shaped structure (along the shear layer) engulfing the unreacted gases in the large induction zone behind the weaker incident shock. This structure is qualitatively similar to that observed by Kiyanda and Higgins [54]. As the triple point propagated downward from Fig. 4a to Fig. 4e, the tongue of dense unburned gases along the turbulent shear layer was noticeably enlarged, and finally pinched off from the main front as a separate island during the collision of the triple point with the bottom wall, as can be seen in Fig. 4f. In the meantime, a transverse detonation of intense luminosity, as documented numerically by Gamezo et al. [55] and experimentally by Bhattacharjee et al. [51], was generated and then rapidly burned the unreacted gases behind the upper incident shock (see frames from Fig. 4f to Fig. 4j). Such formation mechanism of the transverse detonation in a relatively large unburned induction zone through the reflection of the transverse wave from the wall is similar to that observed recently by Xiao and Radulescu [1] in the mixture of $2H_2/O_2/7Ar$. Starting in Fig. 4f, the unreacted gas island was observed to be fully consumed within eight frames, i.e., 103 $\mu$s, which is the timescale of the detonation cell. Nevertheless, the propagation speeds of incident shocks (close to the bottom curved wall) from Fig. 4a to Fig. 4e were measured decreasing from $0.55D_{CJ}$ to $0.45D_{CJ}$, whose average is $0.5D_{CJ}$. Correspondingly, its post-shock ignition delay (at the average speed of $0.5D_{CJ}$) was calculated to be 10.3 seconds, which is five orders of magnitude longer than the experimentally estimated burn-out time of the unreacted gases processed by this shock. Note that the theoretical ignition time was evaluated using the typical constant-volume explosion method with the detailed San Diego reaction mechanism [56]. Therefore, it is clear that the shock-induced auto-ignition theory alone cannot account for the much faster

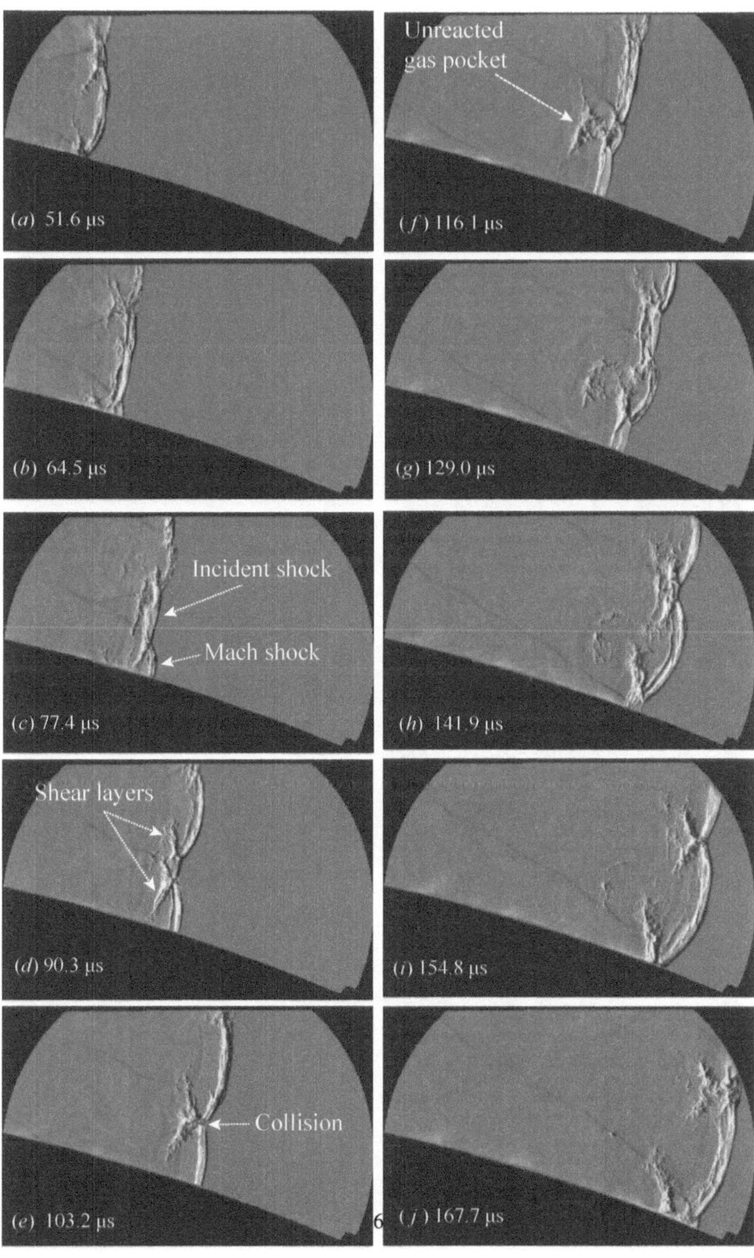
Unreacted
gas pocket
(a) 51.6 µs
(f) 116.1 µs
(b) 64.5 µs
(g) 129.0 µs
Incident shock
Mach shock
(c) 77.4 µs
(h) 141.9 µs
Shear layers
(d) 90.3 µs
(i) 154.8 µs
Collision
(e) 103.2 µs
(j) 167.7 µs

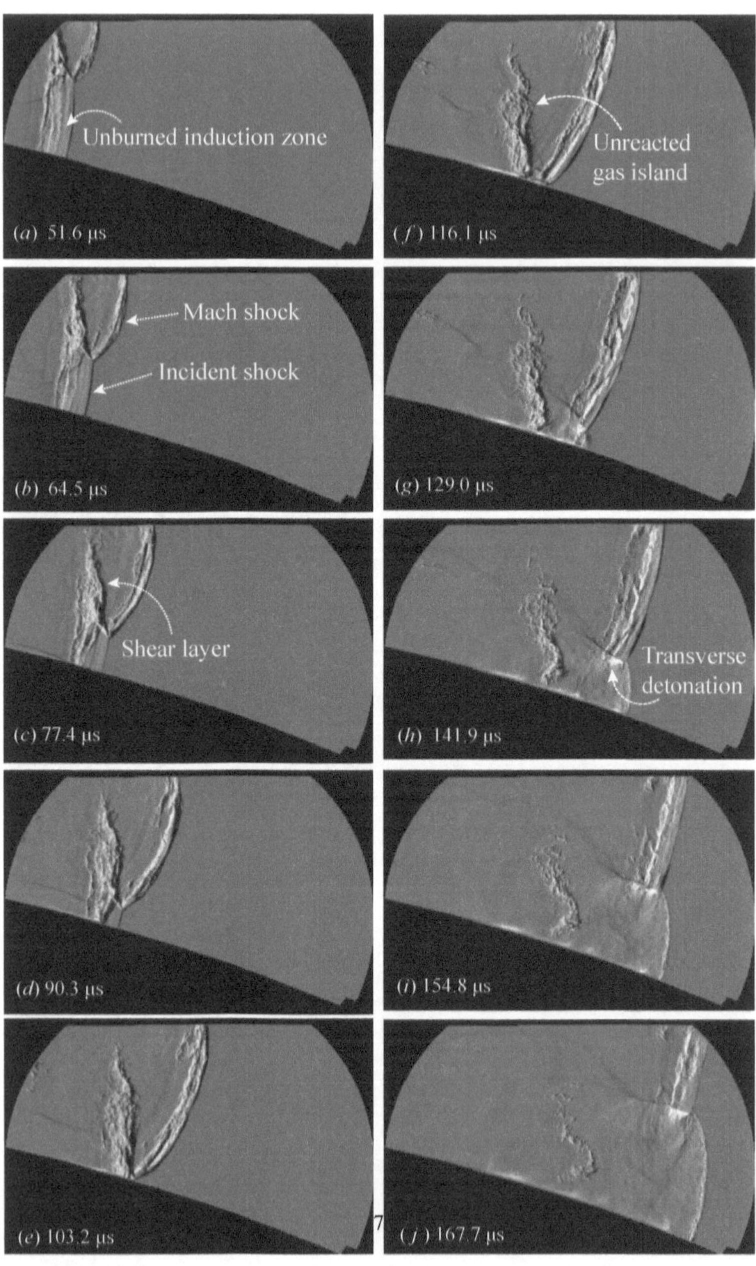


(*a*) 51.6 μs
(*b*) 64.5 μs
(*c*) 77.4 μs
(*d*) 90.3 μs
(*e*) 103.2 μs
(*f*) 116.1 μs
(*g*) 129.0 μs
(*h*) 141.9 μs
(*i*) 154.8 μs
(*j*) 167.7 μs

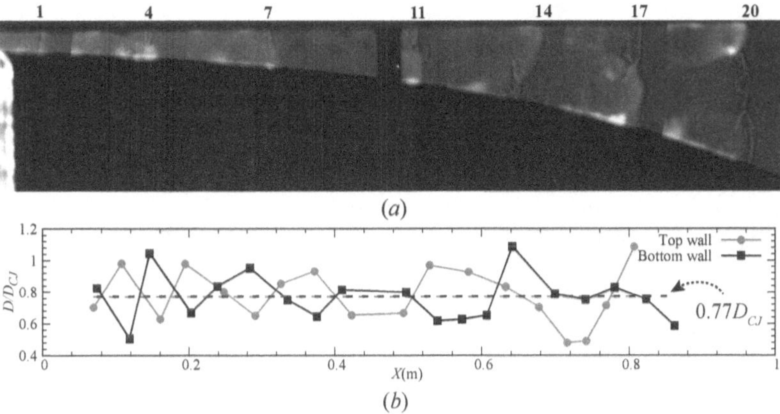

Figure 5: Superposition of detonation fronts ($a$) and the corresponding local speed profiles ($b$) at different time instants along the large ramp in the mixture of $CH_4/2O_2$ at an initial pressure of $p_0 = 13.5$ kPa. Numbers in ($a$) represent the frame sequence. The dashed red and black lines in ($b$) denote the mean propagation speeds over the whole ramp for the top and bottom walls, respectively. Video animations were provided as Supplemental material illustrating the detailed evolution process.

reaction mechanism of the unreacted fuel pockets observed in these highly unstable detonation experiments, as already noted in previous works [48, 49, 53, 54].

In addition, we also obtained the local speed profiles of the near-limit detonation over the whole ramp, as shown in Fig. 5. The locally averaged speeds both along the top and bottom walls (in Fig. 5$b$) were calculated from the corresponding shadowgraph photos (partly superimposed in Fig. 5$a$), with the distance between every two neighboring frames divided by their time interval $\Delta t$. It is clear that the local speed profile in Fig. 5$b$ exhibits periodic fluctuations typical of cellular dynamics, which ranges from the maximum of $1.2\ D_{CJ}$ to the minimum of $0.5\ D_{CJ}$. Despite these significant local variations, detonations in each cellular cycle appeared to propagate at approximately the same mean propagation speed of $0.77D_{CJ}$ along both the top and bottom curved walls. This fact has been detailed recently by Xiao and Radulescu [1] and Radulescu and Borzou [3]. Therefore, it is still meaningful to assume a constant mean propagation speed for the near-limit detonations inside the exponentially diverging channels.

### 3.1.2. Ethylene-oxygen and ethane-oxygen detonations

Figure 6 shows the superimposed schlieren photos of detonation fronts near the end of the large ramp at different instants in mixtures of ethylene-oxygen ($C_2H_4/3O_2$) and ethane-oxygen ($C_2H_6/3.5O_2$), respectively. They are at initial pressures well above the limit of $p_c \approx 1.0$ kPa for $C_2H_4/3O_2$ and $p_c \approx 4.0$ kPa for $C_2H_6/3.5O_2$. The detonation propagated from left towards right. The detonation front acquired a large number of very small-sized cellular structures, and was noticeably curved with a characteristic curvature due to the cross-sectional area divergence. In general, both the $C_2H_4/3O_2$ and $C_2H_6/3.5O_2$ detonation fronts are qualitatively similar, while appear to be less turbulent than the very "rough" ones of $CH_4/2O_2$ in Fig. 2. The $C_2H_4/3O_2$ and $C_2H_6/3.5O_2$ detonations are organized with much thinner characteristic reaction zones.

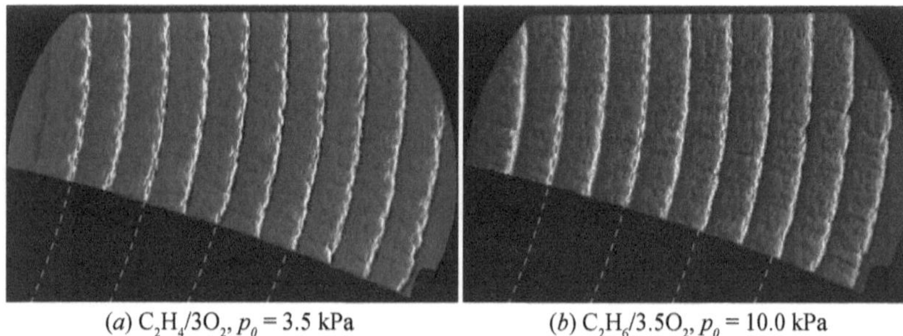

($a$) $C_2H_4/3O_2$, $p_0 = 3.5$ kPa        ($b$) $C_2H_6/3.5O_2$, $p_0 = 10.0$ kPa

Figure 6: Superposition of detonation fronts near the end of the large ramp in mixtures of $C_2H_4/3O_2$ and $C_2H_6/3.5O_2$, respectively, at initial pressures well above the limit. The red lines denote arcs of circles with the expected curvature from the quasi-1D approximation.

On the other hand, the theoretically expected arcs of circles of curvature predicted from the quasi-1D approximation [1, 3] were obtained and compared with the real detonation fronts in experiments. These arcs of circles with the radius of $1/K = 0.46$ m are denoted by the red dashed lines in Fig. 6. The very good agreement between experiments and the theoretical expectations suggests the appropriateness of the quasi-1D assumption for detonations at the macro-scale in the specifically designed exponential geometry, as already demonstrated in detail by Xiao and Radulescu [1] and Radulescu and Borzou [3]. The macro-scale quasi-steady assumption is further confirmed in Fig. 7, illustrating the whole propagation process of $C_2H_4/3O_2$ detonations along the large ramp at a higher initial pressure of $p_0 = 8.3$ kPa. Evidently, detonations propagated with a constant mean front curvature, indicated again by the excellent agreement with the red dashed arcs expected from the quasi-1D approximation. While for the minor deviations recognized near the end, they are due to the limitation of the quasi-1D approximation in designing the exponential ramp [1, 3]. Moreover, the local speed profiles of both the top and bottom curved walls in Fig. 7$b$ show very limited variations within 5% of $D_{CJ}$, indicating detonations travelling at a constant speed. The measured global mean propagation speeds of the top and bottom walls differ within 1% of $D_{CJ}$, i.e., approximately $0.99D_{CJ}$ for the top wall and $0.98D_{CJ}$ for the bottom wall. Therefore, it is reasonable for us to assume a quasi-steady detonation at the macro-scale with a constant mean lateral strain rate in the exponential channel.

In addition, we also observed the propagation characteristics of near-limit detonations in mixtures of ethylene-oxygen ($C_2H_4/3O_2$) and ethane-oxygen ($C_2H_6/3.5O_2$), by reducing the kinetic sensitivity via lowering the initial pressures. In experiments of $C_2H_4/3O_2$, detonations can even self-sustain at very low initial pressures close to 1.0 kPa. Note that for these very low pressures, a driver gas of $C_2H_4/3O_2$ with a higher initial pressure was adopted in the initiation section, which was separated from the second part with a diaphragm shown in Fig. 1. Figures 8 and 9, respectively, illustrate the evolution of single-mode detonations in mixtures of $C_2H_4/3O_2$ and $C_2H_6/3.5O_2$. Compared to $C_2H_4/3O_2$ detonations, the $C_2H_6/3.5O_2$ detonation reaction zone structure appears to be more turbulent. The scenario of generating the tongue-shaped unreacted gas zone along the shear layer is qualitatively similar to that observed in methane-oxygen ($CH_4/2O_2$) detonations in Fig. 4. The significant unreacted gas islands were formed after the reflection of the singe triple point from the wall. These unreacted islands were again observed

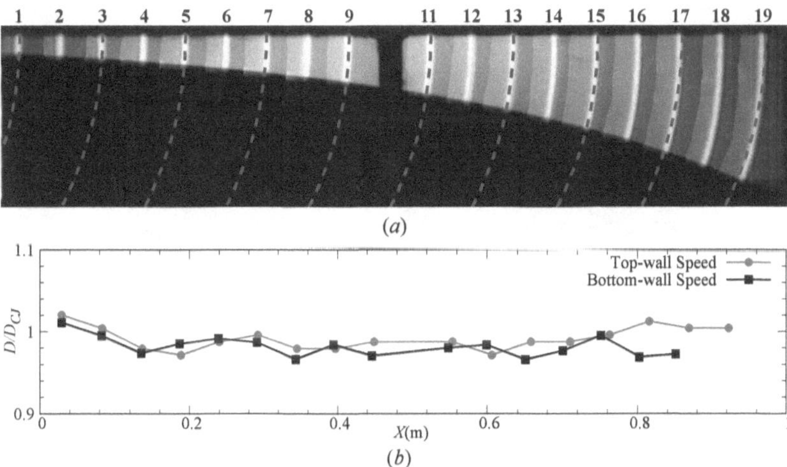

(a)

(b)

Figure 7: Superposition of detonation fronts and the corresponding local speed profiles at different instants along the large ramp in the mixture of $C_2H_4/3O_2$ at an initial pressure of $p_0 = 8.3$ kPa. The red dashed lines in (a) denote arcs of circles with the expected curvature from the quasi-1D approximation.

to be consumed more than four orders of magnitude faster than expected by the adiabatic shock-ignition mechanism.

For $C_2H_6/3.5O_2$ detonations, after collision of the triple point with the bottom wall, a transverse detonation with intense luminosity was generated, as can be seen in Fig. 9$i$ and Fig. 9$j$, while absent in the $C_2H_4/3O_2$ detonation experiment. In Fig. 9, we can also observe the presence of scattered fine-scale unburned pockets (with chemical activity) pinched off from the very spotty main reaction zone, as clearly shown in the sketch of Fig. 10. These unreacted gas pockets were of different sizes and distributed randomly behind the main reaction zone. They were observed to be burned out in a very short time interval, approximately within two or three frames.

### 3.2. The burning rate of unreacted gas pockets

The above near-limit detonation experiments have clearly shown that a significant portion of fresh mixtures are engulfed in tongue-shaped unburned islands along the turbulent shear layers as well as in scattered fine-scale unreacted pockets, and that they react much faster than predicted by the shock-induced auto-ignition mechanism. As such, shock compression alone cannot explain the much faster combustion events observed in these unstable hydrocarbon-oxygen detonation experiments. On the other hand, quite a few studies, e.g., Refs. [48–50, 53, 54, 57], have indicated that these pockets are consumed via surface turbulent flames. The present experiments, as illustrated in Figs. 3, 4, 8, and 9, further confirm this burning mechanism through diffusive flames at very rough pocket boundaries, characteristic of significant hydrodynamic instabilities giving rise to enhanced turbulent mixing between the reacted and unreacted gases along the turbulent shear layers [48–50, 53]. It is thus instructive to estimate the rate at which burning of these unburned fuel pockets occurs. Recently, Maxwell et al. [49] proposed an approximate

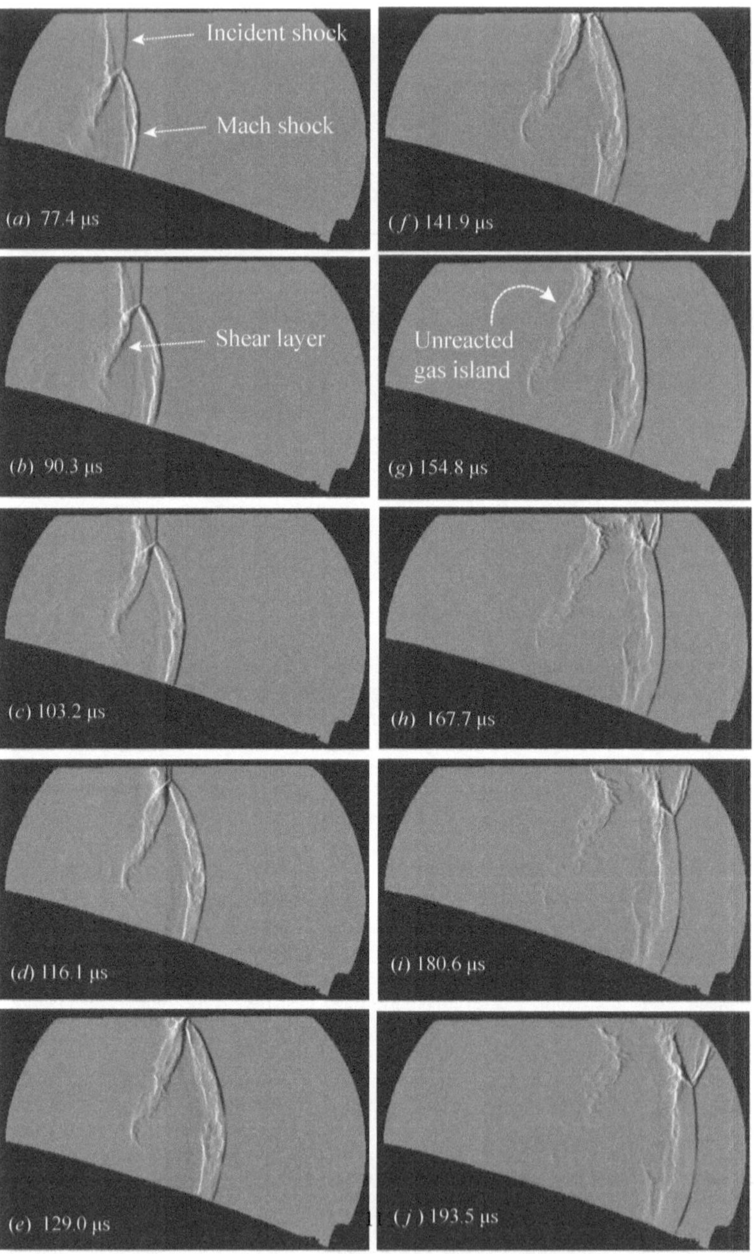

Incident shock
Mach shock
(a) 77.4 µs
Shear layer
(b) 90.3 µs
(c) 103.2 µs
(d) 116.1 µs
(e) 129.0 µs
(f) 141.9 µs
Unreacted
gas island
(g) 154.8 µs
(h) 167.7 µs
(i) 180.6 µs
(j) 193.5 µs

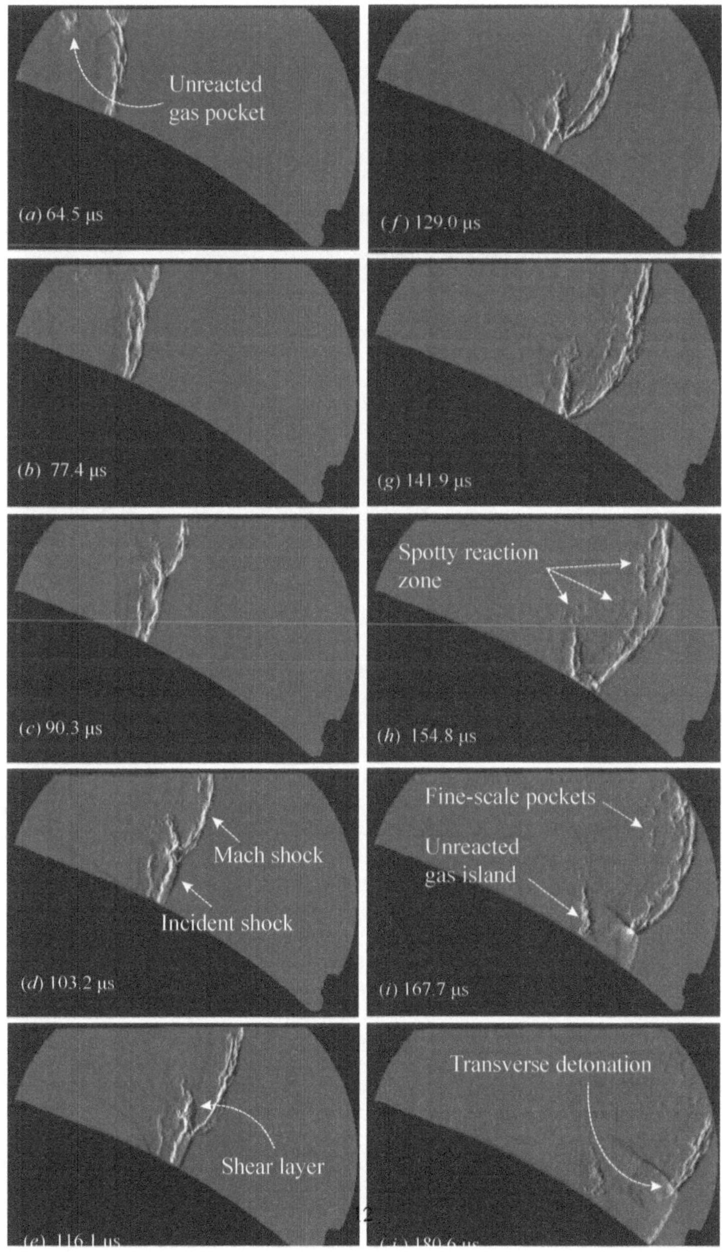
Unreacted
gas pocket
(a) 64.5 μs
(b) 77.4 μs
(c) 90.3 μs
Mach shock
Incident shock
(d) 103.2 μs
Shear layer
(e) 116.1 μs
(f) 129.0 μs
(g) 141.9 μs
Spotty reaction
zone
(h) 154.8 μs
Fine-scale pockets
Unreacted
gas island
(i) 167.7 μs
Transverse detonation
(j) 180.6 μs

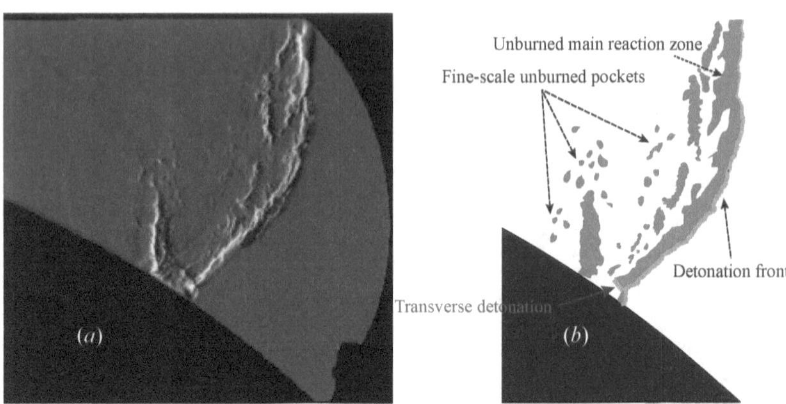



Figure 10: Illustration of the scattered fine-scale unreacted pockets: (a) the spotty reaction zone structure of $C_2H_6/3.5O_2$ detonation in Fig. 9h, and (b) sketch of the main features of the reaction zone.

method for evaluating this turbulent flame speed, by considering an isolated single pocket and measuring its characteristic size and the time interval required for being consumed. Specifically, such turbulent flame speed $S_t$ can be approximately calculated as

$$S_t \approx \frac{L}{\Delta t} \tag{1}$$

where $\Delta t$ is the time interval, which the unreacted fuel pocket takes to be burned out. $L$ is the characteristic length of the pocket, which can be estimated as

$$L = \frac{V}{A_s} \tag{2}$$

where $V$ is the pocket's volume, which is obtained by tracing the pocket shape area, multiplied by the channel width in the third dimension. $A_s$ is the peripheral surface area, estimated with the pocket perimeter multiplied also by the channel width in the third dimension.

Table 1 shows the experimentally evaluated turbulent flame speeds $S_t$ for unburned fuel pockets starting in Fig. 4f, Fig. 8f, and Fig. 9h, respectively. These speeds vary approximately from 30 m/s to 70 m/s, which is smaller than that of $110 \sim 120$ m/s obtained by Maxwell et al. [49] from the $CH_4/2O_2$ detonation experiment in a straight channel. Such discrepancy presumably originates from the much weaker incident shock ($D_{avg} \approx 0.5D_{CJ}$) in the present work resulting in lower post-shock temperatures than those of Maxwell et al. ($D_{avg} \approx 0.7 \sim 0.8D_{CJ}$) [49], as well as in part from the stronger expansion waves behind the leading shock of detonations in the current diverging channel. Moreover, we also estimated the theoretical laminar flame burning velocity $S_L$ and CJ deflagration speed $S_{CJ}$ [58, 59], using the post-shock state of the incident shock that processed the unreacted gas pockets. The results in Table 2 show that, while the experimentally measured turbulent flame burning velocity $S_t$ is smaller than the CJ deflagration speed $S_{CJ}$ by a factor of two to three, it is two to seven times larger than the corresponding laminar flame

Table 1: The estimated turbulent flame burning velocity $S_t$ of the experimentally observed unburned fuel pockets in Figs. 4, 8, and 9.

| Mixture | $p_0$ (kPa) | $V$ (mm$^3$) | $A_s$ (mm$^2$) | $L$ (mm) | $\Delta t$ ($\mu$s) | $S_t$ (m/s) |
|---|---|---|---|---|---|---|
| $CH_4/2O_2$ | 12.7 | 39,000 | 5,900 | 6.6 | 103 | 64 |
| $C_2H_6/3.5O_2$ | 4.6 | 11,000 | 2,900 | 3.8 | 78 | 49 |
| $C_2H_4/3O_2$ | 1.2 | 31,000 | 6,000 | 5.2 | 155 | 34 |

Table 2: The laminar flame burning velocity $S_L$ and CJ deflagration speed $S_{CJ}$, evaluated from the post-shock state forming the unreacted gas pockets by considering the detailed chemistry [56]. Note that $D_{avg}$ is the average speed of the weaker incident shock that processed the unreacted fuel pockets in in Figs. 4, 8, and 9.

| Mixture | $p_0$ (kPa) | $D_{avg}/D_{CJ}$ | $S_L$ (m/s) | $S_{CJ}$ (m/s) | $S_t/S_{CJ}$ | $S_t/S_L$ |
|---|---|---|---|---|---|---|
| $CH_4/2O_2$ | 12.7 | 0.50 | 9.9 | 114.6 | 0.56 | 6.5 |
| $C_2H_6/3.5O_2$ | 4.6 | 0.57 | 12.4 | 113.5 | 0.43 | 4.0 |
| $C_2H_4/3O_2$ | 1.2 | 0.47 | 14.1 | 95.9 | 0.36 | 2.4 |

burning velocity $S_L$. This agrees with the result of $S_t/S_L \approx 6.6 \sim 7.3$ evaluated by Maxwell et al. [49] in their $CH_4/2O_2$ detonation experiments. The decreasing sequence of $S_t/S_L$ from $CH_4/2O_2$ to $C_2H_6/3.5O_2$ and $C_2H_4/3O_2$ also signifies the decreasing level of turbulence effects. It is consistent with the qualitative comparison of the near-limit detonation reaction zone structures in these three mixtures, as shown in Fig. 11, where $CH_4/2O_2$ detonation structures appear to be much rougher while the $C_2H_4/3O_2$ detonation is the opposite.

### 3.3. Velocity deficits

Figure 12 summarizes the detonation mean propagation speeds measured along the top wall over the whole ramp, as a function of initial pressures. As a result of decreasing the mixture kinetic sensitivity through lowering the initial pressures, detonations exhibit larger velocity deficits. Since detonations along the large ramp experience a lower logarithmic area divergence rate than along the small ramp, they show a smaller velocity deficit as well as lower critical pressure, below which detonations fail to self-sustain. Of noteworthy is the relatively large spread of measured points recognized near the limit. This is presumably due to the highly stochastic property of these near-limit detonations, which gives rise to the measured average speeds of fluctuations. Such near-limit stochastic phenomenon has been extensively reported in previous experiments, e.g., see Refs. [1, 60–62].

We further analytically evaluated the uncertainty of the measured average speeds of the near-limit detonations with respect to the physical channel length, as shown in Fig. 13. The abscissa denotes the length of channels, which was normalized by the detonation cell length. The periodic detonation speed evolution in each cell was depicted by an exponential function $D(x)/D_{CJ} = 1.2\exp(-x/1.14)$ with $0 \le x \le 1$, where $x$ is the non-dimensional distance. Such exponential relationship of detonation speeds in a cell can be justified by the linear relation between $D(x)$ and $\frac{dD(x)}{dt}$, as demonstrated by Jackson et al. [63]. The resulting cyclic speed profiles are shown in Fig. 13a, ranging from the maximum of 1.2 $D_{CJ}$ to the minimum of 0.5 $D_{CJ}$, characteristic of the amplitudes observed in the near-limit detonation experiments as shown in Fig. 5b and also detailed in Ref. [1]. The average speed in a complete cycle is $D_{avg} = 0.75 D_{CJ}$. Nevertheless,

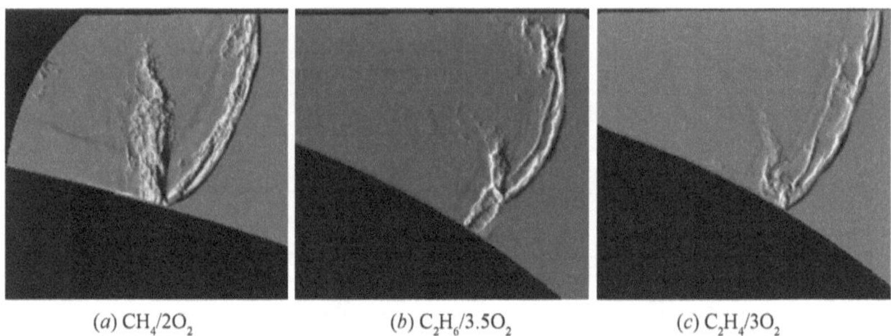



Figure 11: Near-limit structures of (a) CH$_4$/2O$_2$ detonations along the large ramp in Fig. 4e, (b) C$_2$H$_6$/3.5O$_2$ detonations along the small ramp at $p_0$ = 4.5 kPa, and (c) C$_2$H$_4$/3O$_2$ detonations along the small ramp at $p_0$ = 1.8 kPa.

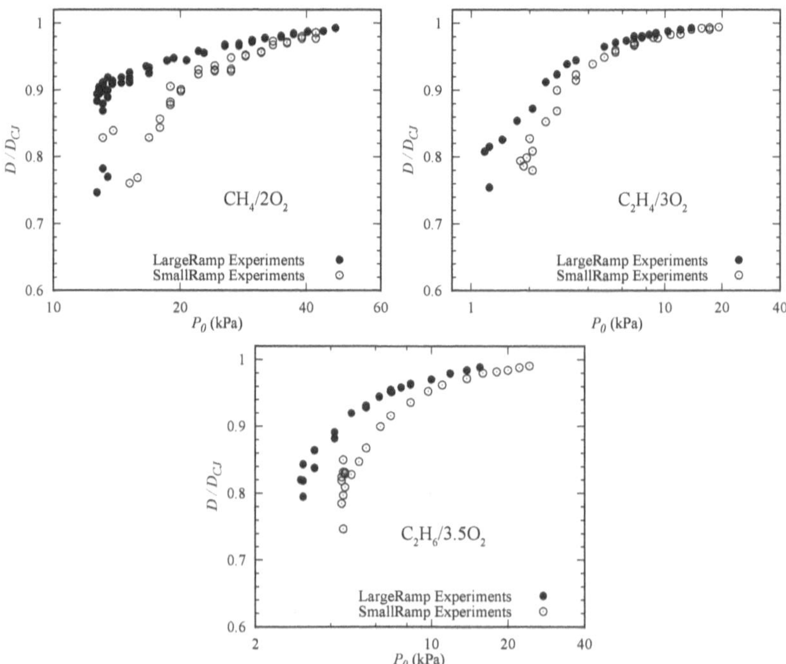



Figure 12: The mean propagation speeds of detonations along the top wall of the whole ramp as a function of initial pressures in mixtures of CH$_4$/2O$_2$, C$_2$H$_4$/3O$_2$, and C$_2$H$_6$/3.5O$_2$, respectively.

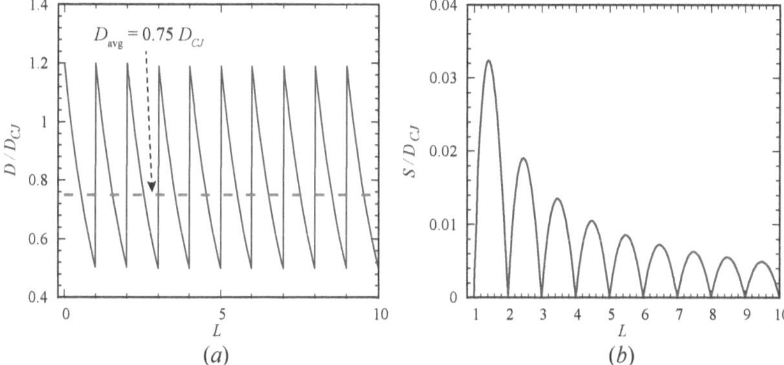

Figure 13: (*a*) periodic detonation speed profiles and (*b*) the standard deviation $S$ of the average speeds with respect to the physical channel length $L$. Note that $L$ was normalized by the characteristic cell length of detonations.

depending on the location, the measured average speeds over a specific channel can vary around $0.75\,D_{CJ}$. As such, we can obtain the standard deviation $S$ of these average speeds from $D_{\text{avg}} = 0.75\,D_{CJ}$ in terms of the channel length, as shown in Fig. 13*b*. Clearly, as the channel length $L$ is longer than one cell, the standard deviation $S$ of the average speed over this channel is within $0.04\,D_{CJ}$. On the other hand, in the present work, the large ramp can host at least four complete cellular cycles for the near-limit detonations, as shown in Fig. 5, while the small ramp can have two. Thus, the uncertainty of the experimentally measured average speeds of these near-limit detonations is within $0.02\,D_{CJ}$, which is negligible. In conclusion, the scatter reported in Fig. 12, for example, for the near-limit detonations, is likely associated with the stochasticity of the detonation dynamics near the limit. Indeed, in diffraction experiments [60, 62], the limit itself was shown to be stochastic.

## 4. Discussion

### 4.1. The experimental $D(\kappa)$ relationships

The total flow divergence experienced by detonations propagating along ramps in this study includes two parts, the one due to the geometrical area divergence of the exponentially diverging channel and the other due to divergence of the flow rendered by the boundary layer growth on the side channel walls [1, 3]. The effective lateral flow divergence rate can thus be expressed as:

$$K_{\text{eff}} = \underbrace{\frac{1}{A}\frac{dA}{dx}}_{K} + \phi_{\text{BL}} \tag{3}$$

where $\phi_{\text{BL}}$ represents the contribution of the boundary layer losses, and $K$ is the known logarithmic area divergence rate of the channel. One can refer to the work of Xiao and Radulescu [1] for the theoretical justification of Eq. (3). Instead of modelling the equivalent flow divergence rate $\phi_{\text{BL}}$ of boundary layers, Radulescu and Borzou [3] proposed to evaluate this loss rate directly from experiments by analytically comparing the experimental data of two ramps, based on the following two assumptions: (1) detonations inside the exponential horn geometry of different divergence rates have the same constant $\phi_{BL}$ since the channel's dimension of the width is unchanged; (2) for the same mixture, it has a unique correlation between the velocity deficit and its loss [2, 64–66]. As a result, the effective rate of total flow divergence $K_{eff}$ can be calibrated by collapsing together the experimental $D/D_{CJ} - K\Delta_i$ curves of both the large and small ramp experiments, by considering the boundary layer effects. As such, the loss rate $\phi_{BL}$ due to boundary layers can be derived. Figure 14 shows the experimentally obtained $D - K$ curves characterizing the relationships between velocity deficits and losses for the mixtures involved in this study. Note that the abscissa is the non-dimensional loss obtained by multiplying the flow divergence rate with the corresponding CJ detonation induction zone length $\Delta_i$ (at the same initial pressure with the experiment), which was calculated by using Shepherd's SDToolbox [67] with the San Diego chemical mechanism [56]. In Fig. 14, the right column is the collapsed $D/D_{CJ} - K_{eff}\Delta_i$ correlation after calibrating the effective flow divergence rate $K_{eff}$ from the $D/D_{CJ} - K\Delta_i$ curves in the corresponding left graph by including the boundary layer effects. The derived constant boundary layer loss rates $\phi_{BL}$ for $CH_4/2O_2$, $C_2H_4/3O_2$, and $C_2H_6/3.5O_2$ are 3.5 m$^{-1}$, 4.0 m$^{-1}$, and 3.0 m$^{-1}$, respectively. It is at present not clear what contributes to these different mean loss rates $\phi_{BL}$ for different mixtures. Since the boundary-layer-induced losses depend on the growth of boundary layers inside the hydrodynamic reaction zone, $\phi_{BL}$ is presumably controlled by the hydrodynamic thickness. More work is required for clarifying the relationship between the boundary layer losses and the hydrodynamic thickness.

One interesting observation of the experimental $D/D_{CJ} - K_{eff}\Delta_i$ curves in Fig. 14 is the extrapolation of the experimentally measured detonation speeds to zero divergence, the experimental data seem to extrapolate to speeds larger than the CJ values by a few percent. This result is the same with the earlier finding of Radulescu and Borzou [3] for the unstable $C_3H_8/5O_2$ detonations. Recently, Damazo and Shepherd [68] also reported the same phenomenon that, when extrapolated to zero losses, the experimental measurements were slightly larger than the CJ predictions by several percent. One can further refer to the review of Fickett and Davis [4] for more discussions and earlier references.

While for comparisons of the experimentally obtained $D - K$ relationships with the generalized ZND model predictions using detailed chemical kinetics and the calibrated global one-step reaction model, as shown in Fig. 14, they will be discussed at length in the following parts of 4.2 and 4.3, respectively.

### 4.2. Comparisons with predictions of ZND detonations subject to lateral strain

To start with, the experimental $D - K$ relationships were compared with those predicted from the generalized ZND model with lateral strain rates [1–3], as denoted by the solid red lines in Fig. 14. The generalized ZND model predictions of quasi-1D detonation dynamics with lateral flow divergence were calculated using the developed custom Python code [3] working under the framework of SDToolbox [67] and Cantera [69]. The chemical kinetics was described by the San Diego reaction mechanism [56]. These calculations were conducted at approximately the medium pressure of the experimental range for each mixture. Normalizing the varied divergence rates $K_{eff}$ by the CJ detonation induction zone length $\Delta_i$ at this medium pressure thus enables us to get the theoretical $D/D_{CJ} - K_{eff}\Delta_i$ curves. The comparisons between experiments and detailed chemistry predictions in Fig. 14 show that while the generalized ZND model appears to predict relatively well the dynamics of $C_2H_4/3O_2$ and $C_2H_6/3.5O_2$ detonations for small velocity deficits (approximately $D/D_{CJ} \geq 0.9$) and flow divergence, it considerably underpredicts the

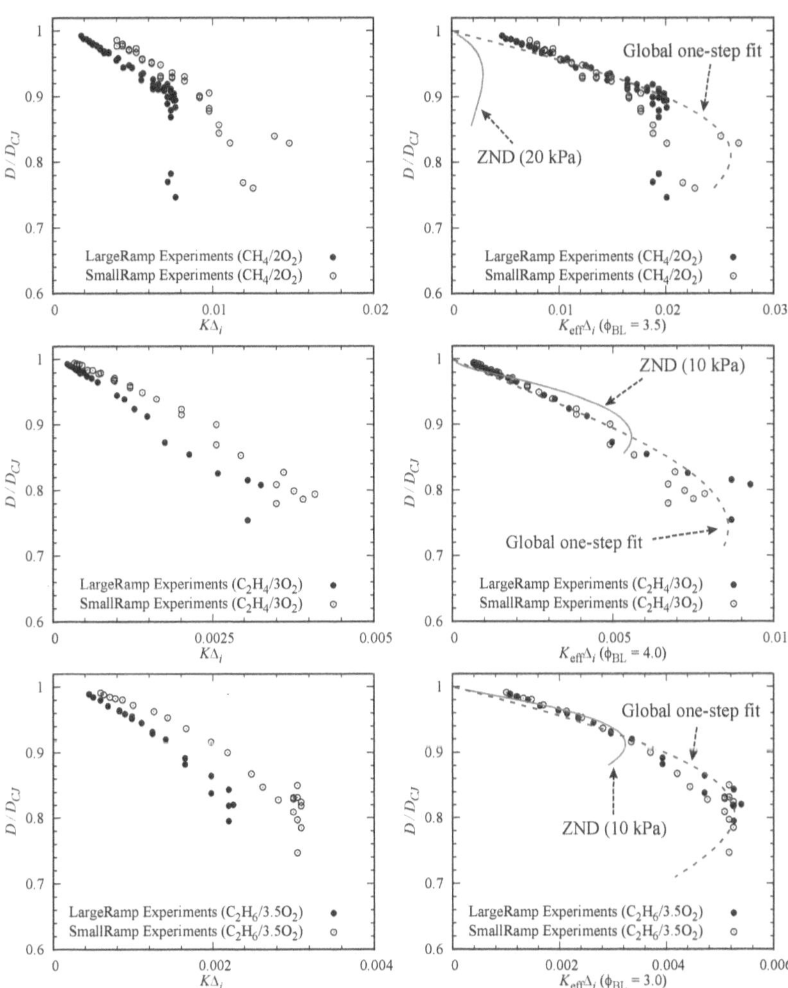

Figure 14: The non-dimensional $D - K$ characteristic relationships obtained experimentally and predicted from the generalized ZND model with lateral strain rates [1–3]. Note that the solid red line denotes the ZND model predictions with the detailed reaction mechanism [56], while the dashed blue line represents predictions using the empirical global one-step reaction model. All these ZND model predictions were calculated at the approximately medium pressure of the experimental range for each mixture.

limiting flow divergence and the maximum velocity deficit. For $CH_4/2O_2$ detonations, the prediction is much worse and completely underpredicts the experiments, where the departure can reach approximately one order of magnitude for the critical flow divergence. Evidently, these comparisons demonstrate that the unstable hydrocarbon-oxygen detonations in experiments can propagate with much larger lateral strain rates and velocity deficits than predicted by the steady ZND model, which concurs with the previous findings of Radulescu et al. [3, 32, 33].

Following the analysis of Short and Sharpe [70], Radulescu [33] proposed the parameter $\chi$ for characterizing the detonation instability level, with detonations in mixtures of higher $\chi$ being more unstable to perturbations in reaction zones than those with lower $\chi$. The mathematical expression of $\chi$ is

$$\chi = \left(\frac{E_a}{RT_s}\right)\left(\frac{t_{ig}}{t_{re}}\right) \tag{4}$$

where $E_a/RT_s$ is the reduced activation energy, $T_s$ the post-shock temperature, $R$ the specific gas constant, and $t_{ig}/t_{re}$ the ratio of ignition to reaction time. Since the initial pressure does not change much the $\chi$ value of the CJ detonation in a specific mixture, we will adopt the initial pressures of the near-limit experiments for calculating the $\chi$ value, for characterizing the detonation instability level of each mixture in the subsequent sections.

Figure 15 quantifies the departure of experiments from the theoretical predictions in terms of the near-limit detonation dynamics, i.e., the critical flow divergence $(K_{\text{eff}}\Delta_i)^*$ and the minimum mean propagation speed $(D/D_{CJ})^*$, for each mixture as a function of the instability level. Note that the relevant data of $H_2/O_2/Ar$, $2C_2H_2/5O_2/21Ar$, and $C_3H_8/5O_2$ were adapted from the experiments already performed by our research group [1, 3]. As the detonation instability $\chi$ value increases from the lowest of $2H_2/O_2/7Ar$ to the highest of $CH_4/2O_2$, as illustrated in Fig. 15$a$, departure of the experimentally obtained maximum lateral strain $(K_{\text{eff}}\Delta_i)^*_{\text{exp}}$ from the ZND model predicted counterpart $(K_{\text{eff}}\Delta_i)^*_{\text{znd}}$ shows a considerably increasing trend. While $(K_{\text{eff}}\Delta_i)^*_{\text{exp}}/(K_{\text{eff}}\Delta_i)^*_{\text{znd}}$ tends to 1 when $\chi$ approximates 0, it can reach one order of magnitude for $CH_4/2O_2$ with $\chi \approx 1500$. Moreover, for the maximum velocity deficits in Fig. 15$b$, the ZND model predicts the increasing $(D/D_{CJ})^*$ with the increase of $\chi$, due to the strong dependence of the critical velocity deficit on the activation energy [2, 65, 66]. The higher activation energy theoretically permits a much smaller critical velocity deficit [2, 65, 66], and these reduced activation energies generally increase with $\chi$, as confirmed in Fig. 16$b$ and Fig. 17$a$, which will be discussed later. On the other hand, the experiments appear to show the opposite trend, with the measured minimum velocities being almost the same or even decreasing slightly with $\chi$. Noteworthy is that the error bar in Fig. 15$b$ represents the minimum global average velocities obtained near the limit for both the large and small ramp experiments. Clearly, for the more stable $2H_2/O_2/7Ar$ detonation with $\chi$ close to 1, the generalized ZND model impressively predicts the same velocity deficit with the experiment, while for the highly unstable $CH_4/2O_2$ detonation, the experimentally measured velocity deficit can be larger than the predicted deficit by half an order of magnitude.

Therefore, as the detonation instability level increases, the degree of departure of both the experimentally estimated critical lateral strain rates and maximum velocity deficits from the ZND model predicted values increases. Here we further propose the departure level (DL) defined by

$$DL = \frac{(K_{\text{eff}}\Delta_i)^*_{\text{exp}}}{(K_{\text{eff}}\Delta_i)^*_{\text{znd}}} \times \frac{(1.0 - D/D_{CJ})^*_{\text{exp}}}{(1.0 - D/D_{CJ})^*_{\text{znd}}} \tag{5}$$

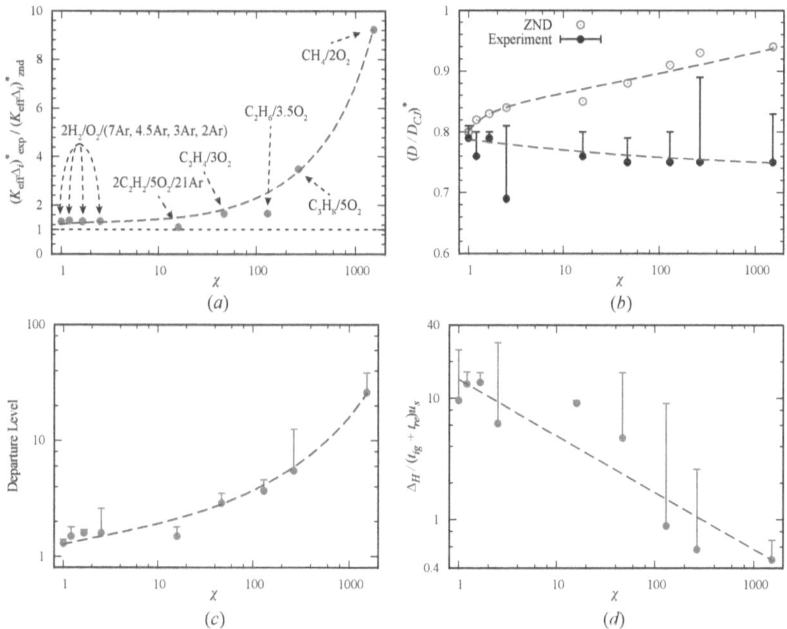

Figure 15: Correlations of near-limit detonation parameters with the instability parameter $\chi$ for different mixtures: (a) ratio of the experimental limiting flow divergence to that predicted by the model, (b) the minimum mean propagation speed near the propagation limit, (c) the departure level (DL) defined by Eq. (5), and (d) ratio of the experimentally estimated reaction zone length $\Delta_H$ to the theoretical one. Note that relevant parameters for the mixtures of $H_2/O_2/Ar$, $2C_2H_2/5O_2/21Ar$, and $C_3H_8/5O_2$ were adapted from the experiments reported by Radulescu et al. [1, 3].

which denotes the departure of the critical flow divergence multiplied by that of the velocity deficit near the propagation limit. Figure 15c demonstrates such departure as a strong function of the instability level $\chi$. Thus, the instability $\chi$ appears to control the level of predictability of detonation dynamics in a specific mixture by the steady ZND model.

In addition, we also evaluated the hydrodynamic reaction zone thickness $\Delta_H$ in the near-limit detonation experiments, by measuring the maximum length between the unreacted gas pockets (immediately before being consumed) and the leading shock. And we then compared it to the theoretical length $(t_{ig} + t_{re})u_s$, where the ignition time $t_{ig}$, reaction time $t_{re}$, and the post-shock particle velocity $u_s$ (in the shock-attached reference) were evaluated behind the shock of the experimentally measured average velocity, using the typical constant-volume explosion method. Compared to the less unstable system of $2H_2/O_2/7Ar$, the highly unstable $CH_4/2O_2$ detonation has a much shorter reaction zone length (observed from experiments) than the theoretical value, as can be clearly seen from Fig. 15d. It thus indicates the existence of additional effects on significantly promoting the combustion in these unstable detonation experiments, thereby decreasing the characteristic reaction length.

*4.3. Empirical global one-step reaction models for the hydrodynamic macroscopic average description of cellular detonation dynamics*

A useful formulation of the detonation problem is seeking a macroscopic average description of the detonation structure, such that relevant empirical models can be developed for capturing the detonation dynamics at macro-scales. The major contribution of the present exponential ramp experiments relies on unambiguously providing a unique curve relating the detonation speed and the hydrodynamic streamline divergence, i.e., the characteristic $D - K$ relationship. Unfortunately, these relations in Fig. 14 differ substantially from the generalized ZND model predictions obtained using detailed chemical kinetics for the mixture. These comparisons show that cellular detonations in experiments generally can propagate under global strain rates much larger than permissible for 1D waves. Indeed, this comes as no surprise and clearly highlights the departure of the global energy release rate in the cellular detonation from that of the steady 1D ZND wave by a significant amount, as clearly indicated by the much shorter ignition time and length scales observed in experiments than the theoretical counterparts.

Despite the ZND model with detailed chemistry failing to correctly capture the response of detonations to the imposed flow divergence, in a practical manner, Radulescu and Borzou proposed using the exponential ramp experiments data to infer an empirical reaction model for the hydrodynamic average description of the macro-scale detonation dynamics [3]. Ironically, the hydrodynamic model is still the ZND model itself, with the fundamental difference that the global energy release rate represents the mean evolution extracted from experiments. The simplest description of the global energy release rate is the one-step reaction model [3], which follows the Arrhenius law as:

$$\omega_P = G \times k_A (1 - Y_P) \exp\left(\frac{E_a}{RT}\right) \tag{6}$$

where $\omega_P$ is the rate of production of product, $Y_P$ the mass fraction of the product, and $k_A$ the typical pre-exponential rate constant. $G$ is the scale factor, which is a tuning parameter. The initial pressure $p_0$, initial density $\rho_0$, and the ZND detonation induction time $t_{ig}$ were adopted as the normalization scales. By doing the same exercise as Radulescu and Borzou [3], the scale factor $G$ and the effective activation energy $E_a/RT_0$ can be fitted such that the ZND model predicted $D - K$ curve matches with experiment.

Table 3 lists all the thermo-chemical parameters required by the empirical reaction model for a global average description of detonation dynamics at macro-scales. $\gamma$ is the post-shock isentropic exponent of the CJ detonation, for correctly recovering the gas compressibility in the reaction zone. The heat release $Q$ was determined from the perfect gas relation [4] recovering the correct detonation Mach number. The real activation energy $E_a/RT_0$ (Detailed) was obtained from the slope of the logarithmic ignition delay with respect to inverse of the post-shock temperature with small perturbations. In these calculations, the detailed kinetic mechanism [56] was applied. Since the maximum velocity deficits of the near-limit detonations are a strong function of the activation energy [2, 65, 66], the effective activation energy $E_a/RT_0$ (Empirical) can thus be readily fitted from these turning points, with their limiting lateral divergence rescaled by tuning the scale factor $G$ in Eq. (6). Of noteworthy is that the scale factor $G$, for the non-dimensional Arrhenius pre-exponential rate constant, actually represents the global reaction rate amplification factor.

Since the resulting one-step reaction models can effectively describe the global rate of energy release in the cellular detonations, as clearly illustrated by the very good agreement between the

Table 3: Thermo-chemical parameters for the empirical one-step reaction model for a global description of cellular detonations, with the reaction rate amplification factor $G$ and the effective activation energy $E_a/RT_0$ fitted from the exponential ramp experiments in Fig. 14.

| Mixture | $p_0$ (kPa) | $\gamma$ | $\dfrac{Q}{RT_0}$ | $\dfrac{E_a}{RT_0}$(Detailed) | $k_A$ | $\dfrac{E_a}{RT_0}$(Empirical) | $G$ |
|---|---|---|---|---|---|---|---|
| $CH_4/2O_2$ | 20.0 | 1.17 | 64.5 | 70.0 | 15.0 | 24 | 8.5 |
| $C_2H_4/3O_2$ | 10.0 | 1.17 | 73.1 | 30.4 | 5.4 | 20 | 2.1 |
| $C_2H_6/3.5O_2$ | 10.0 | 1.15 | 84.5 | 49.9 | 12.2 | 24 | 1.9 |

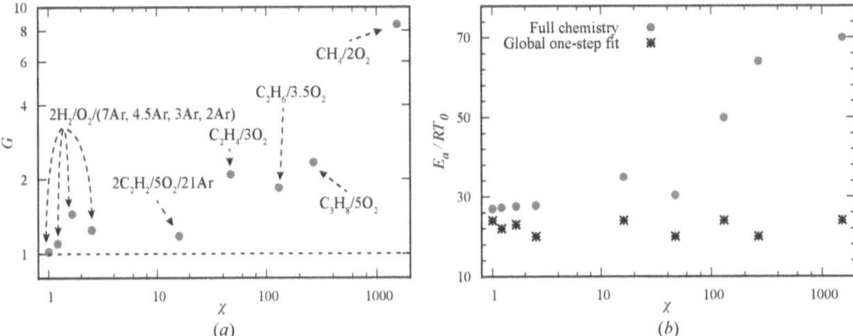


Figure 16: The reaction rate amplification factor $G$ (scale factor for the non-dimensional pre-exponential rate constant) and the global activation energy $E_a/RT_0$, for the empirical one-step reaction model.

experiments and global one-step predictions in Fig. 14, it is worth commenting on the magnitude of the two important fitting coefficients in Table 3. Obviously, the effective activation energy for a global description is much lower than the underlying chemical decomposition. Particularly for $CH_4/2O_2$ detonations, the reduction in energy can reach 66%, signifying a dramatic change in the overall detonation dynamics characteristics. Alongside with these much lower effective activation energies, the global rates of energy release are greatly enhanced, as indicated by the reaction rate amplification factor $G$. It is thus clear that these unstable hydrocarbon-oxygen cellular detonations have a much more enhanced burning mechanism with the substantially suppressed thermal character of ignition, as compared to the laminar ZND wave.

In addition, we also performed the same exercise for the experiments of all the other mixtures done by our research group [1, 3]. Figure 16 shows the two important coefficients, i.e., the amplification factor $G$ and the effective activation energy, in terms of the instability parameter $\chi$ for various mixtures. Clearly, when $\chi$ tends to 0, $G$ becomes 1 and the effective activation energy is the same with the activation energy obtained from detailed chemistry. As a result, the macro-scale detonation dynamics of the more stable systems, such as $2H_2/O_2/7Ar$, can be excellently captured by the ZND model, as detailed by Xiao and Radulescu [1]. With the increase of $\chi$, both the reaction rate amplification effects and reduction of the real activation energy are much more intensified. The instability level $\chi$ appears to correlate well with the qualitative enhancement effects of the burning mechanism.

While an effective method has been devised above for modelling the global mean reaction rate of unstable hydrocarbon-oxygen cellular detonations from experiments, this approach is not entirely satisfactory, since it misses the physical clarity. As such, the modeller has little insight into the cause and effect. Clearly, much more work is required for clarifying the physical relevancy of the present one-step fit for describing the global kinetics of detonation dynamics. Moreover, as the present work and previous ones [48–50, 53, 54, 57] have demonstrated the important roles of turbulent diffusion, future work should also be devoted to bottoms-up approaches [46] for rationalizing and identifying the most promising modelling avenue of the global-scale average exothermicity rate law for the hydrodynamic description of the detonation structure.

On the other hand, despite the limitation of the present one-step fit derived from experiments in terms of physical clarity, our motivation here is to provide simple rate laws from the well-defined experimental benchmark for effectively capturing the macroscopic detonation dynamics for macro-scale engineering applications. For example, one can possibly find a use of these laws in engineering-scale simulations of rotating detonation engines, where quasi-steady and stationary cellular detonation waves may appear and typically experience weak confinements. One can also refer to the report by Radulescu [46] for the usefulness of these laws in predicting the dynamics of detonations, such as initiation and failure. Clearly, much more efforts are required for further validations.

### 4.4. What contributes to enhancement of the global energy release rates in cellular detonations, as compared to a laminar ZND wave?

The above analysis has quantitatively demonstrated that the highly unstable hydrocarbon-oxygen cellular detonations in experiments show a much more enhanced ignition mechanism, as compared to the laminar ZND wave. Thus, the significant question is, *what contributes to enhancement of the global reaction rates in cellular detonations?*

Our clarification starts from the analysis of the relevant time and length scales. When assuming the ignition delay $t_{ig} \sim \exp(E_a/RT_s)$, we can get

$$\frac{t_{ig}}{t_{ig,CJ}} = \exp\left\{\frac{E_a}{RT_{s,CJ}}\left(\frac{T_{s,CJ}}{T_s} - 1\right)\right\} \tag{7}$$

where $T_{s,CJ}$ is the post-shock temperature of the CJ detonation, while $T_s$ is that behind the detonation with velocity deficit. In the limit of strong shock assumption, i.e., $T_{s,CJ}/T_s \approx (D_{CJ}/D)^2$, and high activation energy, by performing a perturbation analysis of Eq. (7), we can then simplify it to

$$\frac{t_{ig}}{t_{ig,CJ}} \cong \exp\left\{\left(2E_a/RT_{s,CJ}\right)(1 - D/D_{CJ})\right\} \tag{8}$$

which is the same with what we have recently obtained for cell sizes varying with respect to velocity deficits [42]. It thus highlights the exponential sensitivity of the ignition time on velocity deficit, which is controlled by the global activation energy.

As the highly unstable detonation has an evidently higher reduced activation energy, as shown in Fig. 17$a$, its response of ignition time to velocity deficit is much stronger. For example, at $D/D_{CJ} = 0.7$, the ignition time of $CH_4/2O_2$ detonations is dramatically increased to approximately three orders of magnitude larger than the CJ detonation, as can be clearly seen from Fig. 17$b$. In cellular detonation experiments, e.g., in Fig. 4, the weaker incident shock can even propagate at a local speed smaller than $0.5D_{CJ}$. Due to the much longer ignition delays behind such weak portions of the leading shock, unreacted gases accumulate and are then pinched away

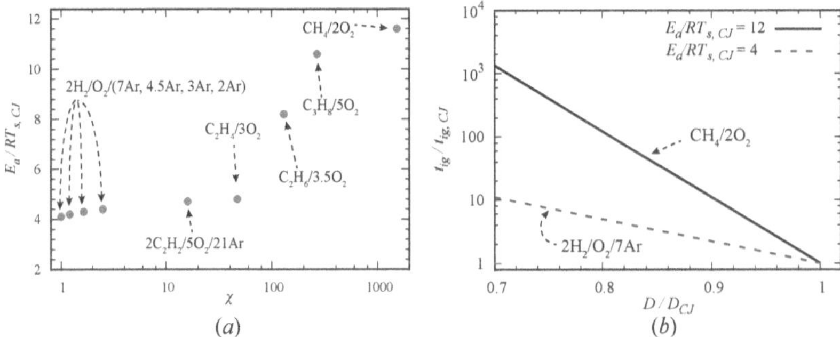



Figure 17: Parameters of (a) the global activation energy $E_a/RT_{s,CJ}$ with respect to instability level $\chi$ for various mixtures, and (b) ignition time ratio predicted from Eq. (8) for $E_a/RT_{s,CJ}$ = 12 (CH$_4$/2O$_2$) and $E_a/RT_{s,CJ}$ = 4 (2H$_2$/O$_2$/7Ar) as a function of velocity deficits.

from the front through expansion, giving rise to unreacted gas pockets distributed in the reaction zone, as clearly demonstrated in Fig. 10. On the other hand, due to the relatively small activation energy, the weakly unstable 2H$_2$/O$_2$/7Ar detonation has a very limited variation of ignition time with respect to velocity deficits, as illustrated in Fig. 17b. As a result, very few significant pockets of unreacted gases have been observed in weakly unstable detonations.

Although these unreacted gas pockets are generated due to the much longer ignition delays behind weak incident shocks, their consumption time has been observed to be comparable to or smaller than the detonation cell timescale, suggesting their much faster reaction rates than the shock-induced auto-ignition mechanism. Moreover, visualization of the evolution of these pockets (see Figs. 4, 8, and 9) further shows that these pockets burn through the very rough pocket boundaries by turbulent surface flames, whose speed is approximately two to seven times larger than the corresponding laminar burning velocity. These much enhanced burning velocities highlight the important roles of turbulent mixing on the unreacted pocket surfaces and also between the reacted and unreacted gases along the turbulent shear layers [48, 49, 57]. Therefore, the much stronger sensitivity of ignition time on velocity deficits associated with the highly unstable cellular detonations results in significant unreacted pockets, whose combustion through diffusive surface flames is considerably promoted by turbulent mixing, thereby substantially suppressing the thermal ignition character and enhancing the global energy release rates.

## 4.5. The $D/D_{CJ} - K_{eff}\lambda$ curves

Finally, we also summarized the $D/D_{CJ} - K_{eff}\lambda$ relationships for various mixtures, as shown in Fig. 18. These relations were obtained by using the already calibrated effective flow divergence rate $K_{eff}$ multiplied by the corresponding cell size $\lambda$. Note that $\lambda$ was obtained by appropriately fitting the cell size data of the corresponding mixture given in the Detonation Database of Caltech [71], while the cell size of 2H$_2$/O$_2$/3Ar detonations (not available in database for the present experimental pressure range) was fitted from data obtained by Zhang et al. [38]. The fitting parameters of $\lambda$ are shown in Table 4. Of noteworthy is that the $D/D_{CJ} - K_{eff}\lambda$ curve of 2H$_2$/O$_2$/3Ar detonations goes beyond that of 2H$_2$/O$_2$/2Ar detonations in Fig. 18, which disagrees with our previous finding that 2H$_2$/O$_2$/3Ar detonations fall between 2H$_2$/O$_2$/2Ar and 2H$_2$/O$_2$/7Ar detonations in terms of the $D/D_{CJ} - K_{\mathrm{eff}}\Delta_i$ relationships [1]. This inconsistency can be clarified due to the fitted cell size of $2H_2/O_2/3Ar$ detonations larger than that of $2H_2/O_2/2Ar$ by a factor of 2, as can be seen from the relations in Table 4. Taking into account the measurement uncertainty of cell sizes [72], these $D/D_{CJ} - K_{\mathrm{eff}}\lambda$ relations appear to follow a relatively good universality, as already found by Nakayama et al. [14, 15]. Such relatively good universality with $\lambda$ implicitly suggests that cell size is also a function of the detonation wave stability, since it is controlled by the reaction zone thickness of unstable detonations.

Table 4: The fitting parameters for the cell size relations of $\lambda = A * p_0^{-B}$ (mm).

| Mixture | $A$ | $B$ | Reference |
|---|---|---|---|
| $CH_4/2O_2$ | 1231.2 | 1.37 | [73–77] |
| $C_2H_4/3O_2$ | 79.4 | 1.10 | [74, 77] |
| $C_2H_6/3.5O_2$ | 261.2 | 1.28 | [77, 78] |
| $C_3H_8/5O_2$ | 190.8 | 1.15 | [77] |
| $2C_2H_2/5O_2/21Ar$ | 240.4 | 1.33 | [43, 79, 80] |
| $2H_2/O_2/2Ar$ | 443.3 | 1.39 | [81, 82] |
| $2H_2/O_2/3Ar$ | 999.1 | 1.38 | [38] |
| $2H_2/O_2/7Ar$ | 1042.8 | 1.40 | [43, 82] |

$p_0$ is the initial pressure in the dimensional unit of kPa.

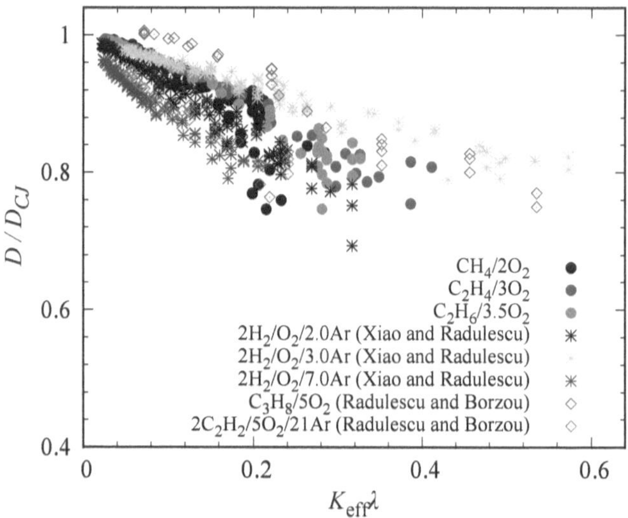


Figure 18: The $D/D_{CJ} - K_{\mathrm{eff}}\lambda$ relationships for various mixtures that have been done by the authors' group [1, 3].

## 5. Conclusion

Experiments of detonations in three different hydrocarbon-oxygen mixtures with varied levels of cellular instability, i.e., $CH_4/2O_2$, $C_2H_4/3O_2$, and $C_2H_6/3.5O_2$, were conducted inside exponentially diverging channels. Visualization of the low-initial-pressure detonation reaction zone structures shows that these unstable detonations are characterized by the presence of significant unreacted gas pockets and turbulent spotty reaction zones. The turbulent flame burning velocity of these pockets was evaluated, and found to be approximately 2 to 7 times larger than the corresponding laminar flame burning velocity, while smaller than the CJ deflagration speed by a factor of 2 to 3. Also, these experimentally visualized unreacted fuel pockets can be consumed more than four orders of magnitude faster than expected by the shock-induced ignition mechanism, thus confirming the previous observations highlighting the burning mechanism of these pockets through turbulent surface flames in unstable detonations [48, 49, 53, 54].

The characteristic $D - K$ relationships between the mean propagation speeds and dimensionless flow divergence were also obtained for each mixture, and compared with the generalized ZND model predictions of quasi-1D detonation dynamics in the presence of lateral strain using detailed chemical kinetics. Generally, these hydrocarbon-oxygen detonations in experiments were found to propagate under much larger limiting global divergence rates with higher maximum velocity deficits than predicted by the steady ZND model. It was also found that the detonation instability level $\chi$ appears to control the level of departure between experiments and the ZND model predictions in terms of the near-limit detonation dynamics. The relatively good universality of the $D/D_{CJ} - K_{\mathrm{eff}}\lambda$ relations implicitly suggests the dependence of cell size on the detonation wave stability.

By adopting the same method as Radulescu and Borzou [3], the empirical global one-step reaction models were developed from the data of experiments in exponentially diverging channels for the hydrodynamic macroscopic average description of cellular detonation dynamics. The significance of the two important empirical coefficients, i.e., the reaction rate amplification factor $G$ and the effective activation energy, was commented upon. The considerably increased global reaction rates indicated by $G$ and much lower effective activation energies than the underlying chemistry decomposition highlight a much more enhanced burning mechanism of the highly unstable detonations through substantially suppressing the thermal character of the ignition, as compared to the laminar ZND wave. The substantially enhanced global rates of energy release thus results in a much shorter hydrodynamic reaction zone length observed in experiments than the theory.

Finally, through analysis of the relevant time and length scales, the ignition time was found to follow the exponential dependence on the velocity deficit, which is controlled by the global activation energy. The much stronger sensitivity of the characteristic ignition time scales on velocity deficits associated with the highly unstable cellular detonations results in significant unreacted gas pockets. The burning of these pockets is further enhanced by turbulent mixing on flame surfaces and also between the reacted gases and unreacted gases along the turbulent shear layers, as already noted by Radulescu et al. [48] and Maxwell et al. [49]. As such, the overall thermal ignition character is suppressed and rates of global energy release are greatly enhanced. As a result, the laminar steady ZND model is not sufficient for capturing these unstable waves.

# Chapter 3

# Numerical Investigation on 2D Cellular Detonations with Lateral Strain Rates

## 3.1 Overview

The previous chapter has systematically investigated the dynamics of detonations with a constant mean flow divergence, in exponential geometries. Characteristic relationships have been obtained and compared with the first-principles predictions of quasi-1D detonation dynamics in the presence of lateral strain rates. However, more realistically, how these lateral strain rates impact the unsteady 2D cellular detonation dynamics still remains unknown. This is because in experiments, it is very hard to isolate the lateral strain rate for evaluating its effect on the unsteady 2D detonation cell dynamics.

On the other hand, for detonations in a narrow channel, as shown in Fig. 3.1, the boundary layer losses have been well known to significantly impact the detonation dynamics, i.e., decreasing the mean propagation speeds and increasing their propagation limit pressures as well as the cell sizes [1]. Nevertheless, the already developed models are mainly on 1D detonations in the presence of wall losses, and no efforts have been made in modelling realistically the effects of the boundary layer losses on unsteady 2D cellular detonations. Moreover, most of the relevant multi-dimensional numerical simulations (e.g., see Refs. [77, 78] ) only consider the ideal detonations without accounting for the losses, presumably due to the fact that directly resolving the boundary layers in DNS computations is a great challenge and computationally prohibitive. For example, the recent works of Tsuboi et al. [79], Chinnayya et al. [51], and Sow et al. [52] proposed that directly resolving the viscous boundary layer requires the resolution of at least 160 points to 320 points per half reaction zone length, for the weakly unstable detonation. In the work of Tsuboi et al. [79], they numerically simulated both the effects of turbulent and laminar boundary layers on the galloping detonation phenomenon in a 2D narrow channel. While the attempts of Chinnayya et al. [51] and Sow et al. [52] have convincingly demonstrated

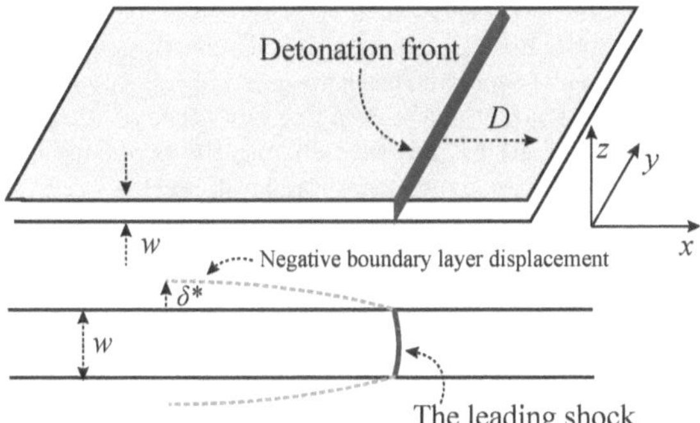



Figure 3.1: Sketch of detonations subject to boundary layer losses in a narrow channel.

the mechanism of boundary layers acting as a mass sink in diverging the flow inside the reaction zones, unfortunately, insights into the effect of these losses on the detonation cellular dynamics were not provided. Therefore, the present work aims to address this problem by proposing a novel quasi-2D formulation of detonations in narrow channels, by modelling the boundary layer losses induced from the third dimension (i.e., $z$ direction in Fig. 3.1) into the 2D reactive Euler equations as a source term. As such, effects of these losses on unsteady 2D cellular detonation dynamics can be readily computed. Corresponding experiments in a narrow channel are performed for the model validation. Note that since the narrow channel width or depth $w$ in the present work is much smaller than the dimension of $y$ direction (by approximately one order of magnitude), the boundary layer losses from $y$ direction is thus negligible compared to that from $z$ direction.

## 3.2 Effect of Boundary Layer Losses on 2D Detonation Cellular Structures

**Status:** This paper has been accepted for publication in *Proceedings of the Combustion Institute*.

This study proposes a novel quasi-2D formulation of the reactive Euler equations accounting for the boundary layer effects from the third dimension, for modelling detonations in narrow channels with significant wall losses. The boundary-layer-induced lateral strain rate is evaluated by applying Mirels' shock-induced laminar boundary layer displacement thickness relation. Corresponding experiments of low-initial-pressure $2H_2/O_2/7Ar$ detonations were performed in a narrow channel for the model validation. The results showed that simulations with the calibrated two-step chain-branching reaction model are in excellent agreement with the experiments in terms of the cellular dynamics, with the Mirels' boundary layer constant modified by a factor of 2. Also, the influence of boundary layer losses on cell dynamics and detonation cellular structures is demonstrated. The present work finally highlights the importance of reporting the corresponding velocity deficits when measuring the detonation cell size, since it can vary up to one order of magnitude! As compared to the 3D DNS computations by directly resolving the boundary layers, this proposed quasi-2D approach cannot only greatly save computational resources, but also permit for quantifying the effect of boundary layer losses on 2D cellular detonations very conveniently.

**Notice:** The supplementary materials of experimental schlieren photos, calculations of the non-dimensional $K_i$ and $K_r$ for the validated two-step reaction model, and also the resolution study, can be found in detail in Appendix A, B and C of this **book**, respectively.

# Effect of Boundary Layer Losses on 2D Detonation Cellular Structures

## 1. Introduction

Detonation structures in narrow channels are usually observed to exhibit a two-dimensional (2D) structure (e.g., see Refs. [1–3]). Nevertheless, they tend to propagate with large velocity deficits due to losses originating from the side walls [4, 5]. The cell sizes are also reported to be much larger in narrower channels, due presumably to increased reaction zone lengths [6–8], and the propagation limits (i.e., the critical initial pressures, below which detonations fail to propagate) are impacted as well (e.g., see Refs. [9–12]).

Previous works focused on modelling the wall losses of detonations in 1D [13–17]. One approach, introduced by Zel'dovich [13], was to model the wall losses with volume-averaged friction and heat loss terms. Another, due to Fay [14], accounted for the negative displacement thickness of the boundary layer, whose effect appeared as a source term of mass flow sink in the governing equations for the inviscid core (i.e., the undisturbed free stream flow). Despite these efforts in modelling 1D detonations, in a realistic way, the effect of these wall losses on the dynamics of 2D cellular detonations in narrow channels remains unknown.

On the other hand, the recent works of Tsuboi et al. [18], Chinnayya et al. [19], and Sow et al. [20] have showed that directly resolving the viscous boundary layers by the Navier-Stokes (NS) equations requires remarkably high resolution, which makes the computations considerably expensive. Moreover, for these NS calculations, the other difficulty lies in quantifying the wall loss effects on detonation cellular structures.

In the present study, we adapt Mirels' technique [21] by accounting for the wall-boundary-layer-induced loss in a 2D formulation of the problem. This permits us to readily compute the dynamics of unsteady 2D cellular detonations with a supplemental lateral loss.

The communication first reports experiments in a narrow channel, with variation of the channel width ($w$) and cell size ($\lambda$) ratios, i.e., $w/\lambda$. We then formulate the governing equations with a lateral loss, which is evaluated from Mirels' boundary layer theory [21]. Comparisons between experiments and simulations follow. Finally, effects of the boundary layer losses on dynamics of the unsteady 2D cellular detonations are explored.

## 2. Experiments in a Narrow Channel

### 2.1. Experimental details

The experiments were performed in a 3.4-m-long thin rectangular aluminium channel with an internal height and width of 203 mm and 19 mm, respectively, as described in detail elsewhere [2]. The shock tube comprises three parts, i.e., the detonation initiation section, the propagation section, and the test section (about 1.0 m in length). The mixture was ignited in the first section by a high voltage igniter, and mesh wires were inserted in this part for promoting the detonation formation. The detonation evolution process was visualized in the test section, and its mean propagation speed over the whole test part was obtained by one 113B24 and five 113B27 piezoelectric PCB pressure sensors using the time-of-arrival method. The presently investigated mixture is the very regular stoichiometric hydrogen-oxygen diluted with 70% argon (i.e., $2H_2/O_2/7Ar$). Since the mixture of $2H_2/O_2/7Ar$ has low reactive sensitivity, a more reactive driver gas of stoichiometric ethylene-oxygen (i.e., $C_2H_4/3O_2$) was used in the initiation section, which was separated from the propagation section with a diaphragm. For visualization, a Z-type schlieren setup [2] was utilized with a light source of 360 W. The resolution of the high-speed camera was $384 \times 288$ px$^2$ with the framing rate of 77481 fps (about 12.9 $\mu s$ for each interval).

### 2.2. Results

Figure 1 shows the schlieren photos of the detonation reaction zone structures, at varied initial pressures ranging from 10.3 kPa to 3.1 kPa. By decreasing the initial pressure for reducing the kinetic sensitivity of the mixture, detonations can be clearly observed to propagate with considerably enlarged cellular structures, with the velocity deficits increased up to 20% ~ 30% of the ideal CJ detonation speed. At a relatively high initial pressure in Fig. 1a, where $w/\lambda$ is about 0.5, one can observe possible 3D-like effects of the overall detonation structure by the presence of duplicate features that do not overlap in the schlieren image. This adds to the difficulty in studying its dynamics under the present resolution. With the decrease of $w/\lambda$ to about 0.25 in Fig. 1b, the detonation structure is qualitatively similar. The appearance of double Mach stems sharing the same triple point indicates that detonation is still relatively unstable in this condition. When $w/\lambda$ is further reduced to less than 0.1, as shown in Fig. 1c and Fig. 1d, detonations become perfectly planar and essentially two-dimensional for the investigation. At these low initial pressures, detonations are organized with relatively large unburned induction

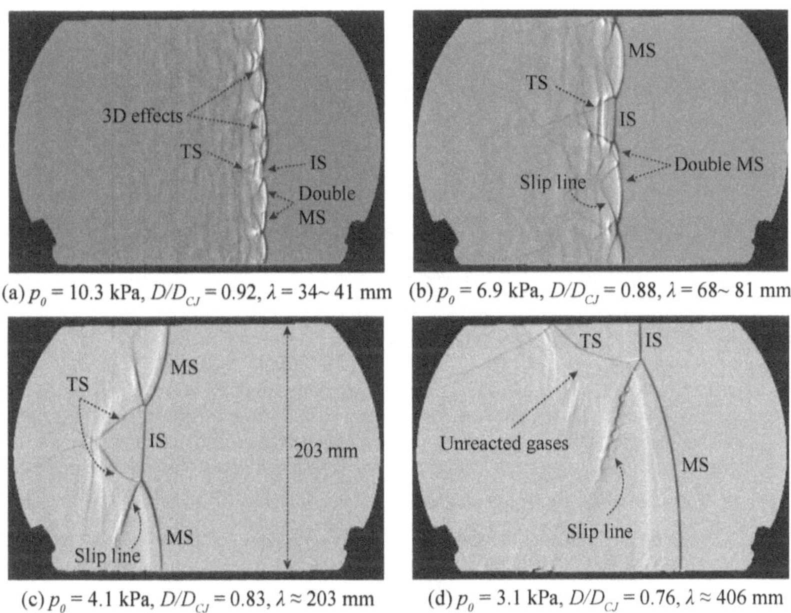



(a) $p_0 = 10.3$ kPa, $D/D_{CJ} = 0.92$, $\lambda = 34\sim 41$ mm



(b) $p_0 = 6.9$ kPa, $D/D_{CJ} = 0.88$, $\lambda = 68\sim 81$ mm



(c) $p_0 = 4.1$ kPa, $D/D_{CJ} = 0.83$, $\lambda \approx 203$ mm



(d) $p_0 = 3.1$ kPa, $D/D_{CJ} = 0.76$, $\lambda \approx 406$ mm

Figure 1: The structure of cellular detonations with varied $w/\lambda$, IS: the incident shock, MS: Mach stem, TS: the transverse shock (video animations as Supplemental material illustrating the evolution process).

zones, as can be observed behind both the leading shock and the transverse wave. The vortex structures characteristic of the Kelvin-Helmholtz instability can also be readily seen along the slip line from Fig. 1d. Since the single head detonation in Fig. 1d experiences a much larger velocity deficit, being more expensive in CFD calculations, we will adopt the condition of Fig. 1c as the main benchmark for subsequent model development and validation of simulations.

## 3. Numerical Simulations

### 3.1. Unsteady quasi-2D formulation

Although the viscous effects are responsible for the boundary layer losses of detonations in thin channels, the present work aims at modelling these effects into the inviscid core flow as a source term. For such transient inviscid reactive core flow behind detonations in a narrow channel, the area-averaged equations of motion across the $z$-direction (i.e., the channel width direction into/out of the page when viewing Fig. 1) can be expressed in the lab frame of reference

$(x, y, t)$ as

$$\frac{\partial \rho}{\partial t} + \frac{\partial(\rho u)}{\partial x} + \frac{\partial(\rho v)}{\partial y} = -\rho \frac{1}{A}\frac{DA}{Dt} \tag{1a}$$

$$\frac{\partial(\rho u)}{\partial t} + \frac{\partial\left(\rho u^2 + p\right)}{\partial x} + \frac{\partial(\rho uv)}{\partial y} = -\rho u \frac{1}{A}\frac{DA}{Dt} \tag{1b}$$

$$\frac{\partial(\rho v)}{\partial t} + \frac{\partial(\rho uv)}{\partial x} + \frac{\partial\left(\rho v^2 + p\right)}{\partial y} = -\rho v \frac{1}{A}\frac{DA}{Dt} \tag{1c}$$

$$\frac{\partial(\rho e)}{\partial t} + \frac{\partial(\rho eu + pu)}{\partial x} + \frac{\partial(\rho ev + pv)}{\partial y} = -\left(\rho e + p\right)\frac{1}{A}\frac{DA}{Dt} - Q\dot{\omega}_R \tag{1d}$$

$$\frac{\partial(\rho Y)}{\partial t} + \frac{\partial(\rho uY)}{\partial x} + \frac{\partial(\rho vY)}{\partial y} = -\rho Y \frac{1}{A}\frac{DA}{Dt} + \dot{\omega}_R \tag{1e}$$

where $\rho, u, v, A, p, Q, Y, \dot{\omega}_R$ denote the mixture density, $x$-direction flow velocity, $y$-direction flow velocity, the cross section area, pressure, heat release, mass fraction and rate of mass production of the single reactant $R$. Note that the present work adopts the reaction of single species with no intermediate radicals, i.e., in the form of $R \rightarrow P$, where $R$ is the only reactant and $P$ the product. The total sensible energy plus the kinetic energy is $e = \frac{p/\rho}{\gamma - 1} + \frac{1}{2}\left(u^2 + v^2\right)$, where a calorically perfect gas with constant specific heats is assumed and $\gamma$ is the ratio of specific heats. The material derivative $DA/Dt$ appearing in the right-hand source term is the cross-sectional area change. Of noteworthy is that one can find the formal derivation of the 1D version by Chesser [22].

For an observer travelling in the frame of reference attached to the leading shock at average speed $D_s$ and following the motion of the fluid, we have the source term in Eq. 1 further expressed as

$$\frac{D}{Dt}(\ln A) = \frac{\partial}{\partial t'}(\ln A) + u'\frac{\partial}{\partial x'}(\ln A) \tag{2}$$

where $A(x, t) = H \times W(x, t)$ is the effective cross section area. $H$ is the fixed channel height of 203 mm in this communication, while $W(x, t)$ is the effective channel width in the $z$-direction. Since the channel height $H$ is much larger than the width $w$ (i.e., $H/w \approx 10$) in the present work, the boundary layer effects in the channel height direction are negligible as compared to those from the channel width direction. $x'$, $t'$, and $u'$ are, respectively, the space and time coordinates and the post-shock flow velocity in the shock-attached frame of reference. Since the motion is pseudo-steady (travelling wave with perturbation), we can neglect $\frac{\partial}{\partial t'}(lnA)$ to the leading order. We thus get

$$\frac{D}{Dt}(\ln A) \approx u'\frac{\partial}{\partial x'}(\ln A) \tag{3}$$

which can be evaluated from Fay's boundary layer theory [14] by using Mirels' compressible laminar boundary layer solutions [21]. The boundary layer displacement thickness $\delta^*(x')$ behind a moving shock is [21]

$$\delta^*\left(x'\right) = K_M \sqrt{\frac{\mu_s x'}{\rho_0 D_s}} \tag{4}$$

where $x'$ is the distance from the shock, $\rho_0$ the density of the flow ahead of the shock, $\mu_s$ the post-shock dynamic viscosity, and $K_M$ the Mirels' constant. One can refer to the recent work of Xiao and Radulescu [3] for details in evaluating $K_M$ for hydrogen-oxygen-argon detonations at varied initial pressures. For $2H_2/O_2/7Ar$ detonations, they found that $K_M \approx 4.0$ [3]. Using this relation, the boundary layer displacement thickness $\delta^*$ in the experiment of Fig. 1c was calculated to be 1.9 mm, with $D_s = 0.8D_{CJ}$ and $x'$ as the hydrodynamic thickness $x_H$ between the leading shock and the sonic surface. Note that $x_H$ was computed from the generalized ZND model with lateral losses [3, 23]. Clearly, the boundary layer is much thinner than the channel width.

Since $W\left(x'\right) = w + 2\delta^*\left(x'\right)$, where $w$ is the physical channel width of 19 mm, we can thus obtain

$$\frac{\partial}{\partial x'}\left(lnA\right) = \frac{2}{w + 2\delta^*\left(x'\right)} \times \frac{d\delta^*\left(x'\right)}{dx'} \tag{5}$$

where $\delta^*\left(x'\right) \ll w$, Eq. 3 then changes to

$$\frac{D}{Dt}\left(\ln A\right) = u'\frac{2}{w}\frac{K_M}{2}\sqrt{\frac{\mu_s}{\rho_0 D_s}}\left(x'\right)^{-0.5} \tag{6}$$

Since Mirels' model assumes that the post-shock state is uniform and steady, to the leading order, we can thus write $x'$ as $x' = u'\left(t - t_s\right)$, where $t_s$ is the time at which the particle crosses the shock. With the mass conservation across the shock $\rho_s u' = \rho_0 D_s$, where $\rho_s$ is the post-shock density, Eq. 6 can be greatly simplified as the following simple expression

$$\frac{D}{Dt}\left(\ln A\right) = \frac{K_M}{w}\sqrt{\frac{v_s}{t - t_s}} \tag{7}$$

where $v_s$ is the post-shock kinematic viscosity, and $t_{elapse} = t - t_s$ is the elapsed time since a particle has passed through the shock front. The shock time $t_s$ is recorded when the shock passes over, and convected with the motion of that particle:

$$\frac{\partial t_s}{\partial t} + \vec{u} \cdot \nabla t_s = 0 \tag{8}$$

### 3.2. Two-step chemistry model

In the experiment of Fig. 1c, we have calculated the post-shock temperatures using the experimentally measured shock speeds along the walls and cell axis. We found that the lowest post-shock temperature is about 900 K, which is still above its cross-over temperature of 800 K to 850 K. Thus, in this study, the two-step chain-branching reaction model [24, 25] will be employed for describing the chemical kinetics. It consists of two components, i.e., a thermally neutral induction zone followed by an exothermic main reaction zone. The transport equations of the induction and reaction variables can be written as:

Table 1: The calibrated non-dimensional parameters for the two-step model from the detailed chemistry.

| $p_0$ (kPa) | $\gamma$ | $E_a/RT_0$ | $Q/RT_0$ | $k_i$ | $k_r$ |
|---|---|---|---|---|---|
| 4.1 | 1.5 | 31.2 | 11.5 | 45.6 | 0.078 |
| 6.9 | 1.5 | 22.8 | 11.8 | 10.6 | 0.11 |
| 10.3 | 1.5 | 24.2 | 12.0 | 12.7 | 0.14 |

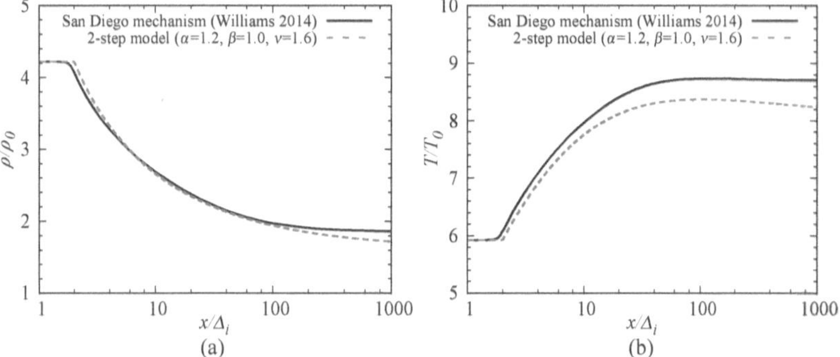

Figure 2: ZND profiles of (a) density and (b) temperature obtained by using the 2-step chemistry model and the detailed chemistry at the initial pressure of 4.1 kPa.

$$\frac{\partial(\rho\lambda_i)}{\partial t} + \frac{\partial(\rho u\lambda_i)}{\partial x} + \frac{\partial(\rho v\lambda_i)}{\partial y} = -\rho\lambda_i\frac{1}{A}\frac{DA}{Dt} - \mathcal{H}(\lambda_i)k_i\rho^{\alpha+1}\exp\left(-\frac{E_a}{RT}\right) \tag{9a}$$

$$\frac{\partial(\rho\lambda_r)}{\partial t} + \frac{\partial(\rho u\lambda_r)}{\partial x} + \frac{\partial(\rho v\lambda_r)}{\partial y} = -\rho\lambda_r\frac{1}{A}\frac{DA}{Dt} - [1 - \mathcal{H}(\lambda_i)]k_r\rho^{\beta+1}\lambda_r^{\nu} \tag{9b}$$

where $\lambda_i$ is the progress variable for the induction zone with a value of 1 in the reactants and 0 at the end of the induction zone, $\lambda_r$ the reaction progress variable with a value of 1 in the unburned zone and 0 in the burned products. $\mathcal{H}(\lambda_i)$ is the Heaviside function given as

$$\mathcal{H}(\lambda_i) = \begin{cases} 0 & \text{if } \lambda_i = 0 \\ 1 & \text{if } \lambda_i > 0 \end{cases} \tag{10}$$

which disables the progress of $\lambda_i$ at the end of the induction zone. $k_i$ and $k_r$ are rate constants, $E_a$ the activation energy controlling the temperature sensitivity of the induction zone duration, $\nu$ is the reaction order, while $\alpha$ and $\beta$ are further empirical reaction order parameters.

Table 1 shows the non-dimensional parameters for the two-step model at three different initial pressures from experiments in Fig. 1. They were calibrated from the detailed chemistry using the San Diego chemical reaction mechanism (Williams) [26], by using Shepherd's Shock and Detonation Toolbox (SDToolbox) [27]. $\gamma$ was the post-shock isentropic exponent of the CJ detonation, while the heat release $Q$ was determined from the perfect gas relation recovering the correct Mach number [28]. The effective activation energy $E_a$ was calculated from the logarithmic derivative of the ignition delay with respect to the inverse of post-shock temperature. Note that the present work adopts the initial state variables $(p_0, \rho_0, T_0)$ and the ZND induction zone length $(\Delta_i)$ as the normalization scales. As such, the rate constant $k_i$ can be directly obtained from Eq. (9a) by scaling the dimensionless induction zone length to unity, while $k_r$ can be determined by recovering the correct induction/reaction time ratio from the detailed chemistry. Finally, $\alpha = 1.2, \beta = 1.0, \nu = 1.6$ were adopted for matching the detailed chemistry ZND structure, as can be easily seen from the density and temperature profiles in Fig. 2. For the minor discrepancies observed near the end of the reaction zone, it is due to the limitation of assuming the constant specific heats in the present work.

### 3.3. Computational details

The non-dimensional governing equations were solved employing the *MG* code, developed by S. Falle of the University of Leeds, which uses a second-order-accurate exact Godunov solver [29] with adaptive mesh refinement. The computational domain height was held constant at the same height of experiments (203 mm in height), i.e., $72\Delta_i$ for $p_0 = 4.1$ kPa, $116\Delta_i$ for $p_0 = 6.9$ kPa, and $182\Delta_i$ for $p_0 = 10.3$ kPa. The domain length varied from $3000\Delta_i$ to $5000\Delta_i$. The detonation propagated from left to the right, with reflective boundary conditions imposed to the top and bottom sides, and zero-gradient boundary conditions applied to the left and right ends. The computations were started using a ZND profile placed $300\Delta_i$ in length from the left boundary. An initial density disturbance zone of $4\Delta_i$ was added ahead of the initial ZND solution for accelerating the evolution to cellular detonations. This density perturbation method is the same as that of Maxwell et al. [30], which is give by

$$\rho(x, y, t = 0) = \begin{cases} \text{ZND solution} & \text{if } x < 0 \\ 1.25 - 0.5n & \text{if } 0 \leq x \leq 4 \\ 1 & \text{otherwise} \end{cases} \tag{11}$$

where $n$ is a random real number from 0 to 1. As for the numerical resolution, 5 levels of mesh refinement were adopted with the coarsest and finest grid sizes of $1/2\Delta_i$ and $1/16\Delta_i$, respectively. Since the reaction zone length of the simulated cases in this work is in the order of $100\Delta_i$ to $1000\Delta_i$, as demonstrated in Fig. 2, such resolution is adequate for obtaining reliable results. This has been verified by a resolution test using a higher level of mesh refinement for calculating the CJ detonation without loss, at the initial pressure of 4.1 kPa. We found that the final stable cell size does not change. In simulations, we ran all the cases for long enough time until we have obtained at least 10 repeated stable cycles of the detonation structures. Due to the large domain size, each case running with 100 to 200 cores in parallel in *Cedar* of Compute Canada requires about one week to complete. More than 20 cases were involved in this study.

### 3.4. Results and discussion

### 3.4.1. Comparison with experiments

Figure 3 shows the numerically tracked maximum energy release rates for detonations with different losses, at the initial pressure of 4.1 kPa. This corresponds to the open shutter photograph in experiments. From the simulated results, it can be observed that the detonation cell size becomes larger as a result of increasing the Mirels' constant $K_M$, i.e., increasing the magnitude of boundary layer losses. As the ideal CJ detonation ($K_M = 0$) has four stable cells across

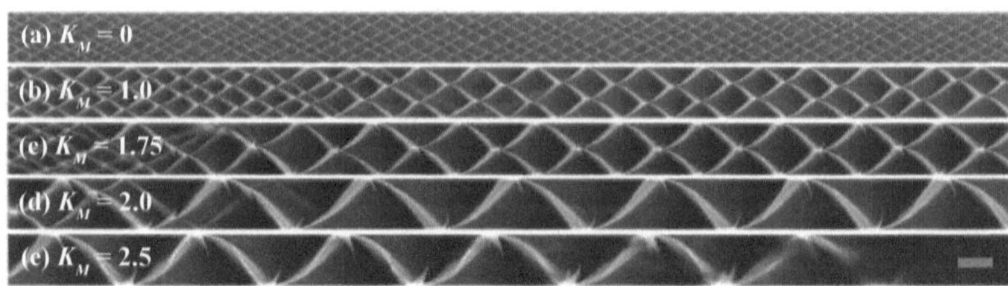

Figure 3: The recorded maximum energy release rates of detonations at the initial pressure of 4.1 kPa with varied $K_M$. Their cell size and mean propagation speeds are: (a) $D/D_{CJ}$ = 1.0, $\lambda$ = 51 mm, (b) $D/D_{CJ}$ = 0.90, $\lambda$ = 136 mm, (c) $D/D_{CJ}$ = 0.85, $\lambda$ = 203 mm, (d) $D/D_{CJ}$ = 0.81, $\lambda \approx$ 406 mm , and (e) detonation failure. In the experiment, $D/D_{CJ}$ = 0.83, $\lambda$ = 203 mm . Note that the red symbol represents $50\Delta_i$ in length, and the length of the shown domain is $1500\Delta_i$.

the channel height, it can only accommodate a single-head detonation for $K_M$ = 2.0. When $K_M$ is further increased to 2.5, the single-head detonation finally failed, as can be seen from Fig. 3e. According to the calculations performed by Xiao and Radulescu [3], the theoretical Mirels' constant for 2H$_2$/O$_2$/7Ar detonations is supposed to be $K_M \approx$ 4.0. However, the present simulations show that $K_M$ = 1.75 can very well recover the experiment (in Fig. 1c) in terms of the cell size and the velocity deficit. This also occurs for cases at other initial pressures of 6.9 kPa and 10.3 kPa, respectively, as shown in Fig. 4. While $K_M$ = 4.0 results in a larger cell size and velocity deficit, $K_M \approx$ 2.5 appears to be able to correctly recover them, when compared to the experiments from Fig. 1a and Fig. 1b. Such discrepancy from the theoretically computed $K_M$ presumably originates from Mirels' assumption of the uniform and steady state behind the shock. For detonations, significant gradients of pressure, temperature, and velocity exist. Particularly, the flow acceleration (in the shock-attached reference) from being subsonic behind the detonation front to sonic in the reaction zone can contribute to thinned boundary layers, as already noted by Chinnayya et al. [19]. In their 2D viscous simulations, they also found that the thickness of the computed boundary layers behind detonations is approximately half of that predicted by the uniform steady boundary layer theory. Moreover, the theoretical calculations of $K_M$ assume the leading shock of the ideal CJ detonation speed, while the 2D simulations have detonations of significant velocity deficits. Thus, it results in a smaller $K_M$ than expected by theory. Future work should be devoted to refine the model to account for these non-idealities. Nevertheless, it is quite satisfying that the model works within a factor of 2, which can also be absorbed by uncertainties in chemical kinetics [31].

Besides the velocity deficit and cell size, the experimentally visualized qualitative features of the detonation structure, as well as its cellular dynamics can be very well reproduced by the simulations, as shown in Fig. 5. Figure 6 also shows the quantitative agreement in temporal velocity evolution at a single cell, when compared to experiments. This suggests the robustness of the proposed quasi-2D formulation as well as the two-step chemistry model in simulating the real detonations in experiments.

### 3.4.2. Dynamics of detonations with different losses

The effect of boundary layer losses on detonation dynamics is shown in Fig. 7, in terms of the normalized speed with respect to the position in a cell. More than 60 points were sampled in a cell for each case, and the local time-average speed was obtained with a running time average

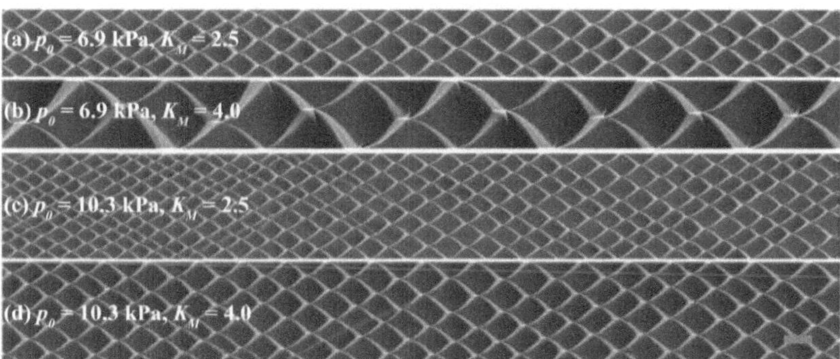



Figure 4: The recorded maximum energy release rates of detonations at the initial pressure of 6.9 kPa and 10.3 kPa, respectively. Their cell size and mean propagation speeds are: (a) $D/D_{CJ} = 0.87$, $\lambda = 81$ mm, (b) $D/D_{CJ} = 0.78$, $\lambda = 203$ mm, (c) $D/D_{CJ} = 0.91$, $\lambda = 41$ mm, (d) $D/D_{CJ} = 0.87$, $\lambda \approx 58$ mm. Note that the red symbol represents $50\Delta_i$ in length, and the length of the shown domain is $1500\Delta_i$.

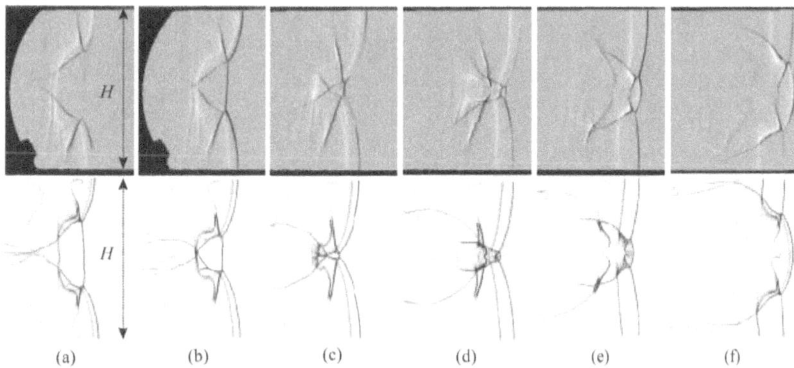



Figure 5: Comparisons of the gradient of the density (bottom column) from the quasi-2D simulation ( $p_0 = 4.1$ kPa, $K_M = 1.75$) with the schlieren photos (top column) from experiments at the initial pressure of 4.1 kPa. $H$ is the channel height of 203 mm.

of 5 neighbouring points. Evidently, the presence of wall losses modifies the cellular dynamics. Compared to the CJ case, the cases with losses have a larger deviation from the larger maximum velocity at the beginning of the cell to the smaller minimum one before the collision of triple points. Whether it is the losses that directly modify the flow velocity inside the cellular structure, or the increased activation energy due to velocity deficits that results in such larger fluctuation, requires further confirmation.

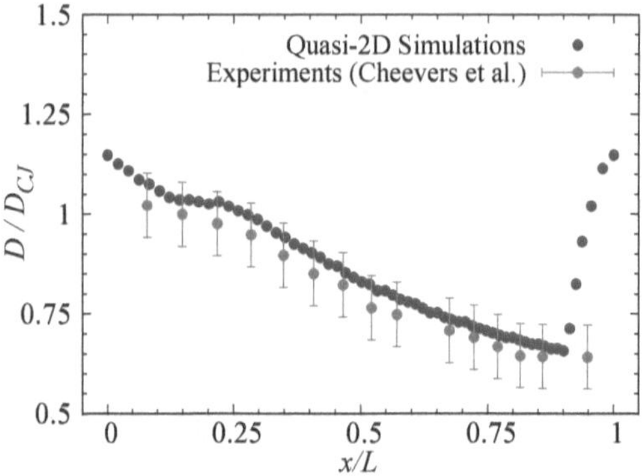

Figure 6: The evolution of detonation speeds in a cell at the initial pressure of 4.1 kPa. Note that the experimental speeds in a cell were reconstructed by Cheevers et al. [32]. The numerical data are the local time-average speed obtained with a running time average of 5 neighbouring points from more than 60 sampling points in a stable cell, for the case of $p_0 = 4.1$ kPa with $K_M = 1.75$. $L$ is the cell length.

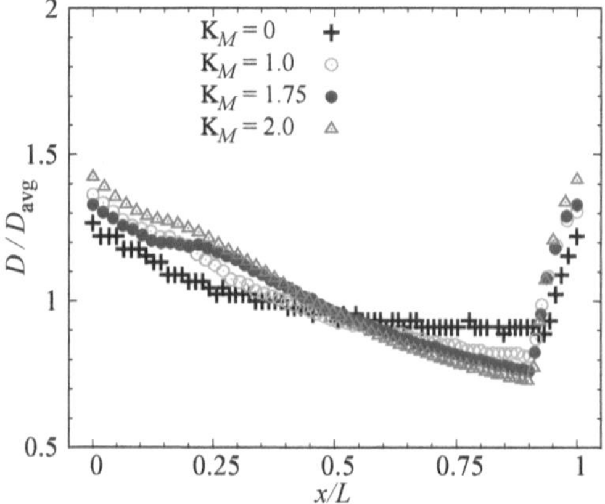

Figure 7: Local time-average speed (along the cell axis) of detonations with different losses, at the initial pressure of 4.1 kPa. Note that the detonation speed is normalized by the mean propagation speed $D_{avg}$ in a cell, and $L$ is the cell length.

### 3.4.3. The $D/D_{CJ} - K_M$ relationships

The variation of the velocity deficit with respect to the Mirels' constant $K_M$ can be better appreciated from Fig. 8. The theoretical predictions were obtained by solving the generalized ZND model with lateral flow divergence [23], using the present two-step reaction model. Clearly, the ZND model underpredicts the velocity deficit obtained from the quasi-2D simulations. As the constant $K_M$ increases to the propagation limit, such discrepancy becomes more significant. These results are consistent with those reported previously by Sow et al. [16] and Reynaud et al.

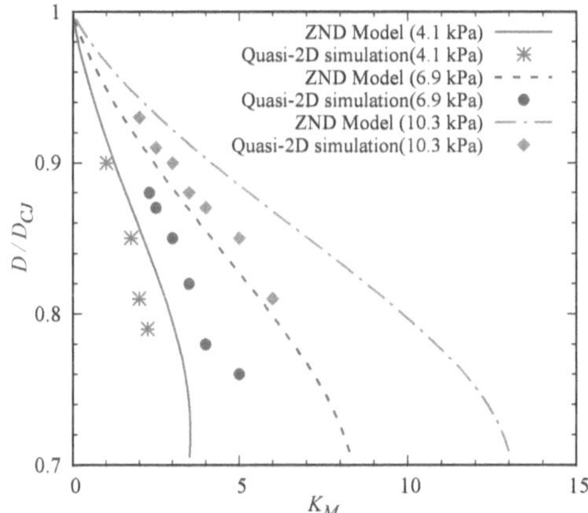

Figure 8: Comparisons of the quasi-2D simulation results with the steady 1D ZND model in terms of the $D/D_{CJ} - K_M$ relationships.

[33], who also found the underprediction of the velocity deficit from the unsteady simulations by the analytical model for relatively regular detonations.

### 3.4.4. The $\lambda/\lambda_{CJ} - D/D_{CJ}$ relationships

We have further obtained the dimensionless relationships in terms of the detonation cell size ($\lambda$) and the characteristic induction zone length ($\Delta$) with respect to the velocity deficit, as shown in Fig. 9. $\lambda_{CJ}$ and $\Delta_{CJ}$ represent the corresponding length scales of the ideal CJ detonation. The induction length $\Delta$ was obtained through zero-dimensional constant-volume combustion calculations with the post-shock velocity (in the shock-attached frame of reference) multiplying by the time to the peak thermicity, as proposed by Shepherd [34]. The detailed chemistry mechanism was utilized in these calculations. The excellent agreement between the $\lambda/\lambda_{CJ}$ and $\Delta/\Delta_{CJ}$ correlations suggest that the increase in cell size due to wall losses is still controlled by the increase in the induction zone length as a result of the velocity deficits.

Assuming that the cell size varies with the induction zone thickness [34] and that the ignition delay $t_{ig} \sim \exp(E_a/RT_s)$, we can thus get

$$\frac{\lambda}{\lambda_{CJ}} \approx \frac{\Delta}{\Delta_{CJ}} = \frac{u'_s}{u'_{s,CJ}} \times \frac{t_{ig}}{t_{ig,CJ}}$$

$$\approx \left(\frac{u'_s}{u'_0}\right)\left(\frac{u'_0}{u'_{s,CJ}}\right)\exp\left\{\frac{E_a}{RT_{s,CJ}}\left[\frac{T_{s,CJ}}{T_0}\right)\left(\frac{T_0}{T_s}\right) - 1\right]\right\} \tag{12}$$

where $u'_s$ and $u'_0$ are the flow velocity behind and ahead of the shock in the shock-attached frame of reference, respectively, and $T_s$ the post-shock temperature. The terms in brackets can be readily evaluated from the shock-jump equations. In the limit of strong shock and high activation energy, we can further simplify Eq. (12) to $\lambda/\lambda_{CJ} \approx \exp\left\{\left(2E_a/RT_{s,CJ}\right)\left(1 - D/D_{CJ}\right)\right\}$. It thus

11

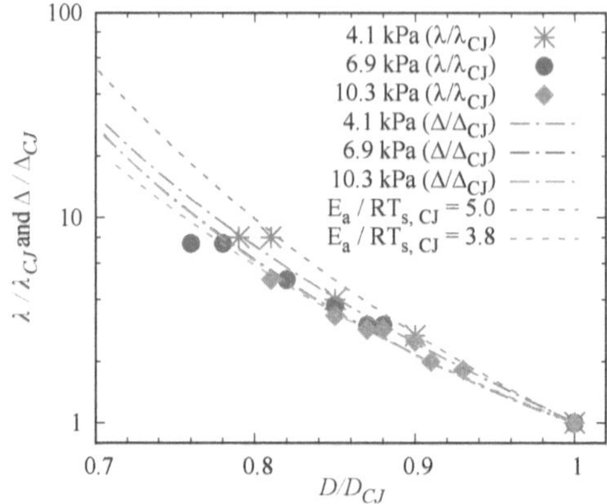

Figure 9: The dimensionless cell size ($\lambda/\lambda_{CJ}$) and characteristic induction zone length ($\Delta/\Delta_{CJ}$) as a function of the velocity deficit. The symbols are from the quasi-2D simulations. The predictions (using Eq. (12)) with $E_a/RT_{s,CJ} = 5.0$ and $E_a/RT_{s,CJ} = 3.8$ correspond to initial pressures of 4.1 kPa and 10.3 kPa, respectively. Note that the real-chemistry constant-volume combustion calculations ($\Delta/\Delta_{CJ}$) of 6.9 kPa (denoted by the purple dashed line) and 10.3 kPa (denoted by the green dashed line) follow almost the same curve.

highlights the exponential sensitivity of the cell size and induction zone length on velocity deficit, which is controlled by the global activation energy. This generalizes Desbordes' observations for overdriven detonations [35]. The results in Fig. 9 show that the $\lambda/\lambda_{CJ} - D/D_{CJ}$ correlations can be well predicted by the simple expression (12). The sensitivity of cell size on velocity deficits also highlights the importance of providing the detonation speed when reporting experimentally measured cell size, since the variation can be up to an order of magnitude, even for the weakly sensitive mixtures studied here.

## 4. Conclusions

The present study has shown that the dynamics of 2D cellular detonations in narrow channels can be well captured using a quasi-2D approach modelling the lateral boundary layer losses using Mirels' theory. With an appropriate Mirels' constant, $K_M$, deviating by approximately a factor of 2 from the model proposed by Mirels for steady constant pressure boundary layers, the simulations are found in excellent agreement with experiment. Compared to directly resolving the boundary layers of detonations in the present narrow channel experiments by calculating the 3D NS equations, this novel formulation can save the computational cost by up to five orders of magnitude. We have also shown that the cellular cycle dynamics is also affected by the losses, which yield larger velocity fluctuations and more rapid decay rates of the lead shock. Finally, the increase in cell size with increasing velocity deficit follows the Arrhenius dependence of ignition delay on the temperature of an equivalent steady shock, in spite of the cellular dynamics, generalizing previous observations of Desbordes for overdriven detonations in generally regular mixtures.

# Chapter 4

# Summary and Conclusions

## 4.1  Summary

The present book demonstrated the excellent predictability of the $H_2/O_2/Ar$ detonation dynamics by the generalized ZND model with lateral strain rates. Such excellent agreement was clarified due to the much longer thermally insensitive reaction zone lengths of argon-diluted hydrogen-oxygen detonations, as compared to the characteristic induction zone lengths. While for the unstable hydrocarbon-oxygen detonations, as compared to their ZND counterparts, they were found to be characterized by the substantially enhanced global rates of energy release with the notably decreased effective activation energies for ignition. Such significant enhancement of the burning mechanism, resulting from the intensified auto-ignition assisted by the turbulent diffusive burning of the unreacted gas pockets, gives rise to the substantial departure observed between the experimentally obtained detonation dynamics and the steady ZND model predictions. The degree of departure has been further shown to correlate well with the detonation instability $\chi$ parameter.

Note that recent works [71, 80–82] have suggested that the isentropic exponent $\gamma$ may also be a controlling parameter impacting the detonation instability, with detonations in gas of low specific heats ratios being more unstable and irregular. Future work should clarify the influence of the isentropic exponent on detonation predictability as well.

On the other hand, the quasi-2D simulations with the proposed numerical model can very well reproduce the $2H_2/O_2/7Ar$ detonation experiments both in terms of cellular dynamics and velocity deficits, provided the theoretical Mirels' constant $K_M$ is reduced by a factor of 2. The presence of boundary layer losses can modify the detonation cellular cycle dynamics, with larger velocity fluctuations and more rapid decay rates of the lead shock. The increase in cell size with increasing velocity deficit was found to follow the Arrhenius dependence of ignition delay on the temperature of an equivalent steady shock, in spite of the cellular dynamics, generalizing previous observations of Desbordes for overdriven detonations in generally regular mixtures.

## 4.2   Conclusions

In conclusion, this study first verified the previous finding that only a particular class of detonations can be well approximated by the steady ZND model. Their predictability strongly depends on the detonation instability. The more unstable detonations have much higher global energy release rates with the considerably suppressed thermal character of ignition than those of the ZND model. Implications of such globally enhanced burning mechanism highlight the important role of diffusive processes, involved in the turbulent burning of the significant unreacted gas pockets. The presence of lateral strain rates can modify the detonation cellular dynamics as well as the cell size, whose increase follows the exponential dependence on the increase of velocity deficits, with the sensitivity controlled by the global activation energy.

## 4.3   Contributions of the Work

### 4.3.1   Original knowledge

Contributions of the present **book** to original knowledge include:

- This study first demonstrated the excellent predictability of the $H_2/O_2/Ar$ detonation dynamics by the generalized ZND model with lateral strain rates using detailed chemical kinetics. It thus verified the previous findings that only a particular class of detonations can be well approximated by the steady ZND model. A novel clarification for this excellent agreement was proposed.

- The link of departures between experiments and the steady ZND model predictions to detonation instability has been clearly established. The role of diffusive processes, involved in the burning of unreacted gas pockets, in affecting detonation limits was examined. Empirical global reaction rate laws have been proposed for effectively capturing the mean evolution dynamics of unstable detonations. The usefulness of these calibrated simple reaction models can be demonstrated in CFD applications, such as predicting the critical energy for detonation initiation phenomena.

- A novel quasi-2D approach was proposed for numerically modelling the effect of boundary layer losses on 2D detonation cellular dynamics in narrow channels. With the theoretical Mirels' constant $K_M$ modified by a factor of 2, the experiments can be very well reproduced by simulations using the resulting quasi-2D formulation. As compared to directly resolving the boundary layers using 3D Navier-Stokes (NS) simulations, this method can greatly reduce the computational cost by approximately five orders of magnitude for the benchmark case in section 3.2.

- Finally, this work further showed that detonation cellular cycle dynamics can be modified by the presence of boundary layer losses, yielding larger velocity fluctuations and more rapid decay rates of the lead shock. The exponential sensitivity of detonation cell sizes on velocity deficits, controlled by the global activation energy, highlights the importance of providing the detonation speed when reporting experimentally measured cell sizes.

## 4.3.2  Engineering applications

The significance of the present book can be further manifested in engineering applications, from the prevention of industrial explosions in the safety field to the development of detonation-based propulsion systems.

From the aspect of the safety field, the major contribution of the present work relies on providing the predictability of the detonation behaviors. With the semi-empirical reaction rate laws obtained in this study, one can have an effective model capable of predicting the detonation dynamics, particularly the detonation limits. Given the knowledge of the propagation limits, one can thus prescribe corresponding instructions in the safety design of industrial pipelines as well as the choice of fuel conditions, with which the initiation and propagation of explosions can be prevented. This also applies to the design of detonation arresters.

Likewise, in the field of engine applications, the semi-empirical reaction models can also be adopted in CFD applications for solving the engineering problems of RDEs. Employing these already calibrated simple rate laws can greatly reduce the computational cost, as compared to using the detailed chemical mechanisms, and in the meanwhile provide the capability for capturing the macro-scale detonation dynamics. Besides, in the framework of DSD, the experimentally obtained characteristic $D(\kappa)$ relationships are essential in the prediction of detonation dynamics in RDEs. Therefore, these useful implications of the present work add to its importance in engineering applications.